Land System Reform in China Since the 1980s

This book is part of the Peter Lang Regional Studies list.
Every volume is peer reviewed and meets
the highest quality standards for content and production.

PETER LANG
New York • Bern • Berlin
Brussels • Vienna • Oxford • Warsaw

Shouying Liu

Land System Reform in China Since the 1980s

Translated by Thomas E. Smith

PETER LANG
New York • Bern • Berlin
Brussels • Vienna • Oxford • Warsaw

Library of Congress Cataloging-in-Publication Data

Names: Liu, Shouying, author. | Smith, Thomas E., translator.
Title: Land system reform in China since the 1980s / Shouying Liu,
Thomas E. Smith.
Description: New York: Peter Lang, 2021.
Includes bibliographical references.
Identifiers: LCCN 2020057939 (print) | LCCN 2020057940 (ebook)
ISBN 978-1-4331-8466-6 (hardback) | ISBN 978-1-4331-8467-3 (ebook pdf)
ISBN 978-1-4331-8468-0 (epub)
Subjects: LCSH: Land reform—China. | Land use, Rural—China. | Land
tenure—China. | Urbanization—China.
Classification: LCC HD1333.C6 L581713 2021 (print) | LCC HD1333.C6 (ebook) |
DDC 333.3/151—dc23
LC record available at https://lccn.loc.gov/2020057939
LC ebook record available at https://lccn.loc.gov/2020057940
DOI 10.3726/b17928

Bibliographic information published by **Die Deutsche Nationalbibliothek**.
Die Deutsche Nationalbibliothek lists this publication in the "Deutsche
Nationalbibliografie"; detailed bibliographic data are available
on the Internet at http://dnb.d-nb.de/.

This edition is an authorized translation from the Chinese language edition
直面中国土地问题. Published by arrangement with China Renmin University Press.

Supported by fund for building world-class universities (disciplines) of Renmin University of China.

80 Broad Street, 5th floor, New York, NY 10004
www.peterlang.com

Contents

Introduction

The Fundamental Logic of Land System Reform in China

Land system reform in China has always been a hot topic and, at the same time, a controversial one, upon which consensus is difficult to reach. After the Third Plenary Session of the 18th Central Committee of the Communist Party of China (CPC) (2013) issued its "Decision,"[1] people throughout society generally responded well to that part relating to land system reform, but there were also a few dissenting voices, which is normal. In the face of controversy, the central government determined the principles of land system reform: it set the program, tested it in local pilots, and only then did it apply the reform uniformly, first by enacting the laws, and so on. In my opinion, in cases where social consensus has not yet been fully achieved, the approach whereby policies and systems are formed through pilot programs and the legitimacy of reform is guaranteed through legal amendments is correct in this area.

From a research perspective, there are many contending voices on this issue. Having such great divergences of opinion is a good thing, since it can stimulate us even more to ponder the matter more rationally and comprehensively and subject it to more objective analysis.

One problem I have been thinking about recently is the necessity of carefully clarifying the strategies and pathways of land reform since the Third Plenary Session of the Eleventh Central Committee in order to find a solution to the land issue, that is, we need to reflect on the questions—What is the logic of the reforms to China's land system? What problems has this logic encountered, and does it still work well?

One of my views is that during the past few years, as the reforms have been implemented, there has always been a search for compromises, and by now they may have touched on fundamental issues. If we do not work hard on the root of land issues or find breakthroughs on the deeper problems, then it may be difficult for land system reform to make great progress.

1. The Two Major Constraints on Choice and Change in China's Land System

China's land system reform has been subject to two major constraints: a system constraint and a goal constraint; these two constraints determine the choice and change in the entire land system.

The first is the system constraint. China's land system is a basic system of the state; the collective ownership of rural land and state ownership of urban land are currently the main forms of public ownership. From the reform decision-making perspective, when reforming the system, questions such as how to change it, to which extent should it be changed, and how would alternative system arrangements affect the public ownership system are of greatest concern.

The system constraint is chiefly manifested in three areas.

First, the legitimacy of governance. Because the ruling basis for the CPC is the elimination of private ownership, its insistence on public ownership determines the legitimacy of governance. The CPC took agricultural land from the hands of landlords and distributed it to the peasants—this was the magic weapon that the CPC used to obtain political power. Public ownership of land is also a foundation of its legitimacy in governance. Second, socialism with Chinese characteristics. The "characteristics" are embodied in the leadership of the CPC and in its adherence to public ownership. The system of public landownership is the most "characteristic" systemic arrangement among the Chinese "characteristics." Third, the basic economic system. The "basic" is embodied in public ownership playing a dominant role and different economic sectors developing side by side, and the land system is the most basic system within the basic economic system. The existence of these system constraints basically determines which of the land system reforms are optional and which are not.

The second major constraint is the goal constraint. During the Second Plenary Session of the Seventh Central Committee (1949), Chairman Mao Zedong proposed changing China from an agricultural country to an industrial one. The CPC since 1949 has been working hard toward this goal. The core of this change consists of realizing economic modernization.

Regardless of whether China was in the planned economy period or in the reform period, and regardless of how controversial the issues were or how many difficulties and setbacks were being faced, economic modernization has always been a fundamental consensus and the goal of the ruling party. The Third Plenary Session of the 18th CPC Central Committee put forward what some people call the "fifth modernization"—the modernization of the country's system and capacity for governance, which in fact is the modernization of the system.

Under the old Soviet model, if one wanted to realize the transformation from an agricultural country into an industrial one, one could only choice industrialization led by state based on depriving peasants. This development model determined the basic pattern of interests wherein agriculture served industry, and the countryside served the city. Since the land system arrangement was the main tool that served this goal, change in the land system could not conflict with the orientation of this goal.

Therefore, choice and change in China's land system have always been subject to two constraints: the fact that the land system is the most important form of public ownership, and the fact that the distribution of land interests must serve the goal of changing from an agricultural country to an industrial one. These two constraints have also determined the basic threefold logic of China's land system reform:

First, the ownership system is locked. The system of public ownership in the land system is locked, and the forms of other systems are excluded. The main manifestations of public ownership are the state ownership of urban land and the collective ownership of rural land. These cannot be modified. What reform can do is to explore the forms in which public landownership is realized.

Second, the basic mainline of land system reform makes an issue about the separation between ownership and use rights. While the first level of the land system's structure is ownership, the second level is formation of rights. Since the first level is locked, breakthroughs in reform can only be sought on the second level, that is, by separating the two rights, expanding use rights, putting into full play the incentivizing and stabilizing functions of property rights, and mobilizing land users' enthusiasm in order to improve land use efficiency.

Third, land becomes a tool for development. Economically speaking, land is a derivative demand, when it should be an inductive one. That is, land plays whatever role economic demand wants it to play—but, as we have passed from a planned economy to the period of reform and on to the present, we somehow got things backwards. Throughout our development, land has really played the role of an engine—it became a tool with which to pursue development. For transforming China from an agricultural country to an industrial one, and for achieving the goal of economic modernization, it has been land that fulfills the engine for growth: only if it runs at the very front can it pull the great wheel for economic growth forward.

Three types of land actually played a very important role in Land system reform: one is the farmers' contracted land; the second is rural residential land or homesteads, where the farmers build their houses, and the third is converted farmland, that is, land converted from agricultural use to other uses. In land system reform, since these three types of land are basically locked in the ownership system, the whole process of reform has started from the logic of "separation of two rights" (*liangquan fenli* 两权分离).[2] However, the rights function has been weakened on each levels.

How shall we comprehend this? For farmland, that is, contracted land, the current logic is to maintain collective ownership rights and strengthen farmers' use rights. That is to say, property rights are strengthened after "separation of two rights" (*liangquan fenli* 两权分离), and then the system of ownership is changed so that ownership becomes a legal and nominal ownership. Although the "two rights have been separated," the latter is stronger. This is the logic of reform for farmland.

For rural residential land, things are different. Although there is a separation of collective ownership rights and use rights for rural residential land, as the system has evolved for this type of land, the outcome is that the former has been strengthened, while the latter has become vague. When the homestead system was being built, collective ownership rights became stronger and stronger, while the protections of farmers' use rights, within the institutional framework, were lacking, so that the property function of homesteads for farmers has almost disappeared.

The situation for converted farmland is even worse. Basically, the conversion of farmland consists of stripping farmers of their collective ownership rights and their use rights. When conversion sets into motion the structures of urban state ownership and rural collective ownership, the outcome of the "separation of the two rights" has in actuality become a division between the two systems of ownership. This division leads to the cancellation of both the collective ownership rights and the use rights.

2. Choice and Change in the Farmland System

2.1 Efforts to Reform the Farmland System Have Never Stopped

Institutional choices and changes with regard to farmland have attracted the utmost concern. Before the agricultural land reform of the 1980s took place, the results of such institutional choices after 1949 may be summarized in four sentences:

One: *Private ownership of farmland by landlords was abolished.* This "land reform" （*tu gai* 土改）basically consisted of abolishing private ownership by landlords and replacing it with ownership by peasants, but the private ownership system still remained.

Two: *The farmers' ownership system was replaced with a cooperative ownership system, that is, elementary communes.* During the era of elementary communes (*chu ji she*初级社, 1949–1956), communes were still privately owned, but the commune system replaced the farmer ownership system.

Three: *The cooperative system was replaced with the collective ownership system.* During the Advanced Cooperatives and People's Commune era, the collective ownership system replaced the cooperative system.

Four: *The system of "production (including land) were owned on three levels, i.e., the commune, the production brigade and the production team"* (san ji suo you, dui wei ji chu 三级所有，队为基础), *and the production team was taken as the basic accounting unit*). The three levels meant that one level of ownership belonged to the production team, but in fact there were several other levels of ownership that continued to play a role; however, the production team was still the foundation.

Research on the defects of the collective ownership of farmland had already yielded a conclusion. Since administrative power replaced the function of property rights, the incentives for producers were low, and rewards were incommensurate with the amount of effort they put in, leading to poor production performance. Consequently, reform of collective ownership system took place—initiated at the grassroots level and then implemented with the support and interaction with reformers at the local and central government levels—whereby farm output quotas were established for each household. These reforms were implemented throughout the country, and then institutionalized through law.

Many people now criticize rural reform for having stagnated since the mid-1980s. In my opinion, this criticism is not very appropriate. In fact, the problem is not that there has been no reform—reform has always been going on, and efforts have continued since the mid-1980s. The key lies in the fact that the choice of system has been settled, so that things can only be done this way. That is, the question should be: How can one strengthen use rights without disturbing the ownership system?

First, the later reforms started with the contracting period. The most important thing for farmers as far as property rights are concerned is stabilizing farmers' expectation. That is, they want the contract period to be continuously extended to the point that their relationship with the land is stable.

A 15-year contracting period was set starting in 1984, and in the second round of contracting, it was extended to 30 years (during this period, it was proposed that "an increase in the number of people [in a household] will not entail an expansion of the land, and a decrease in the number of people will not entail a shrinking of the land" (*zeng ren bu zeng di, jian ren bu jian di* 增人不增地，减人不减地), and the experiment conducted in Meitan, Guizhou, was elevated into national policy and law). In 2008, the "over a long term without change" （*chang jiu bu bian* 长久不变） proposal was raised. This institutional arrangement was made to stabilize expectations.

There has always been a lot of controversy about "an increase in the number of people [in a household] will not entail an expansion of the land, and a decrease in the number of people will not entail a shrinking of the land" (*zeng ren bu zeng di, jian ren bu jian di*增人不增地，减人不减地). This was an institutional arrangement devised in 1987 by the former Central Rural Policy Research Office, on the basis of the experiment at Meitan, Guizhou. At a time when it appeared that reforms would not be able to go on further, Mr. Du Runsheng (杜润生) proposed the establishment of a reform pilot zone, and there were three aspects of the proposed land system. One of them, tried in Meitan, Guizhou, was "an increase in the number of people [in a household] will not entail an expansion of the land, and a decrease in the number of people will not entail a shrinking of the land" (*zeng ren bu zeng di, jian ren bu jian di*增人不增地，减人不减地). In fact, this was a fundamental reform of the entire collective ownership system.

One of the most important arrangements in the collective ownership system is membership in a collective economic organization: everyone has a share. Since everyone has a share, then if someone is added, he too should have a share, and if someone is subtracted, then his share should be removed. Therefore, the result of this membership-based arrangement in the system is that land adjustments are made when the number of members increase or decrease. But the result of that, in turn, would be that farmers' expectations would be affected, and their shares of the land would get smaller and smaller.

The pilot project in Meitan meant that the future relationship between farmers and the land would be fixed, and not adjusted for any increase or decrease of members. From that time onward, land would no longer move no matter who holds it.

The proposal of extending the contracting period "over a long term without change" (*chang jiu bu bian*长久不变) was raised in 2008. As a system arrangement, "over a long term without change" (*chang jiu bu bian*长久不变) meant that not adjusting land on the basis of any increase or decrease of collective members turned into a basic system arrangement. What does "over a long term without change" (*chang jiu bu bian*长久不变) mean? The phrase as it originally appeared in the central government documents was "...[to be] implemented over a long term without change, based on a stable subcontracting relationship." This means that people were to be in full correspondence with the land.

Several reform proposals were raised at the same time that "over a long term without change" (*chang jiu bu bian*长久不变) was suggested in 2008. One was to double the 30-year length of the subcontracting term, or extend it to 70 years, the same age as the state ownership. The opposing voices at the time pointed out that even if the contractual period was extended from 30 to 70 years, it would still be a land system with a time limit. Another point of view was that a "permanent tenancy system" (*yong dian zhi* 永佃制) should be implemented. Some people at the time objected to this, saying it was a philosophical concept—nothing is really permanent—so the arrangement was later called "over a long term without change" (*chang jiu bu bian*长久不变).

The institutional arrangement of "over a long term without change" (*chang jiu bu bian*长久不变) really meant changing from a land system with a time limit to a land system with no time limit, so that whoever holds the land now can continue cultivating it for a long time. However, it said nothing about who the land belongs to—that would have been tantamount to breaking the first bottom line. One could only speak of who has the contractual management right of this land.

The second reform after the mid-1980s was to fine-tune property rights. Property rights consist of the rights to use, derive income from, and transfer property. Actually, a great deal of effort has been made in this area in rural land reform. For example, the first reform in this area was to cancel the "unified purchase and sale system" (*tong gou tong xiao* 统购统销) in place during the early reform period; when the income right was given to farmers, the right to income was fine-tuned. In 2004, the Agricultural Tax was abolished, so that all income derived from the land went to farmers. Under the second reform, the principal status of farmers was constantly emphasized, so that property rights were transferred to families and farming households. Farming households are the principals of property rights—this is something that has been emphasized since the 1980s, and it was constantly re-emphasized in the 1990s, because many people thought that the farming household is too small a unit, or that farmers are ignorant and do not understand the market. To maintain the principal status of farming households, the family contracting management system was even defined as a basic operating system and the "household contract responsibility system with the combination of unification and separation" was written into law.

In addition, starting in 1984, farming households have been permitted to transfer land, and since then, the regulations on land transfer have been constantly fine-tuned, and the principles of "on a voluntary basis and in accordance with relevant laws, transfer the land in return for compensation" have been established.

2.2 Outcomes of the Logic of the Reforms "Separating the Two Rights" (liang quan fen li 两权分离)

Obviously, the people who were engaged in designing the system, with its logic separating ownership and use rights, were very painstaking in their efforts. If we can continue going down this path of reform, then we shall do so. But now that it seems there are more and more problems, what have been the outcomes of the logic of the reforms "separation of two rights" *(liang quan fen li 两权分离)*?

The first problem is that the "collective" (*ji ti* 集体) emerges from time to time. The original intentions of the first stage of the reforms were to make use rights real, to let property rights play their role, to let the farmer cultivate the land well, and to let the collective retain ownership in law—provided that the collective does not cause any trouble. However, the facts show that later on, the collective (*ji ti* 集体) does appear occasionally, because the collective ownership system still exists as an important component of public ownership. Sometimes it is the government that lets it pop up. When we conducted research at the grassroots level, we found that the "collective" never keeps to its place, and when it does, nothing can be done, because, in the law, the "collective" is also primary form.

The second problem is that the explanations attached to whatever "things" are grown by the "collective" (*ji ti* 集体) are vague and unclear.

This is a very big issue in rural governance nowadays. Currently, the "collective" (*ji ti* 集体) shows its real strength mainly by being first to act; strongmen within the collective take it upon themselves to convert farmland into land used for construction. This is what happened in Huaxi Village (a wealthy community in Jiangyin, Jiangsu) and some villages surrounding Beijing. The things grown by the "collective" (*ji ti* 集体) nowadays cannot be separated from two institutional genes: one depends on collective land, and the other depends on collective membership.

However, to whom do the things grown on "collective" (*ji ti* 集体) land belong? As the vague, unclear things grown from the mother get bigger and bigger, more and more people get mixed up in them. This problem in rural governance is occurring repeatedly. Incidents like the "Wukan Incident" (2011, in Guangdong) are not just individual cases, because systemic factors are also at play.

The third problem is that the contract-issuing and contracting relationship between collectives and farming conflicts with empowering farmers with land property rights. As long as the collective's ownership of land currently in the farmers' hands does not change, the collective economic organization contracts it to the farmer. This sort of arrangement, in law, is a kind of contractual relationship, that is, an obligatory creditor–debtor relationship. However, the *Real Right Law of the People's Republic of China* (《物权法》) grants farmers land property rights. As the right of the creditor becomes property, how can the contracting and subcontracting relationship under collective ownership be realized?

The fourth problem is that the idea of membership rights has been strengthened, and it hinders the exercise of exclusivity in property rights. Given that ownership by the collective does not change, the outcome of reform is that ownership by the collective has become ownership by members. The central policy intended to sever this tie, but the problem now is that people have internalized their membership.

In farmers' thinking, collective ownership means that everyone has a share. But once "everyone has a share," this generates a very big conflict with two elements of the system mentioned above—"an increase in the number of people [in a household] will not entail an expansion of the land, and a decrease in the number of people will not entail a shrinking of the land" (zeng ren bu zeng di, jian ren bu jian di增人不增地，减人不减地), and, "over a long term without change" (chang jiu bu bian长久不变). A farmer will say, "This isn't collective ownership? Then why is it, when my family grows, we don't get more land? And why is it that, when my neighbor's family shrinks, they don't lose any land?" They will also ask, "Why is it that when the collective's assets increase, it doesn't give me any share?"

Why is this so? It is because the entire concept of membership has strengthened into something deeply rooted in the farmers' thinking. The collective ownership that farmers now recognize is the rights of membership, but the final result of recognizing the rights of membership is that it becomes difficult to realize exclusiveness of property rights.

For example, Nanhai (Nanhai District, Foshan, Guangdong) had always been exploring reform of the collective shareholding system, and in the end it proved very difficult. Initially, when the agricultural land became collective land used for construction, the collective said, "You won't have to cultivate your land any more—we'll convert it to shares, and then everyone will be paid dividends based on their shares, and the land will be collected and managed." This was in fact an extension of the rights of membership to nonagricultural land, and the farmers said yes to it.

However, if this continues, one faces a big problem, and that is equity welfarization. As long as you are a member of a collective, there will always be money to share, and as a result, the assets grown on such a "collective" will never really mature. With the development of time, more and more wealth emerged, the farmers are not at ease. Later on, the people in Nanhai tried to make the equity fixed—that is, existing members of the collective organization received dividends based on their membership, and when people were added to the collective, they must purchase shares if they want to received dividends. But because the new people did not agree with it, finally this approach was eventually done away with. Now they use families as units for the fixed equity: instead of continuously dividing shares within the collective, the families divide things among themselves.

The fifth problem is the combined protection of "contracting rights and management rights"—it is felt that neither is being protected. The *Law of the People's Republic of China on Rural Land Contracting* (《中华人民共和国农村土地承包法》) and the *Real Right Law of the People's Republic of China* (《物权法》) protect farmers' rights to land contracting and management (tu di cheng bao jing ying quan 土地承包经营权). When the contractor and the manager is combined, there is no problem with "rights to land contracting and management" (tu di cheng bao jing ying quan 土地承包经营权) protected as one right.

In the early days of reform, farmers were basically land-holding farmer. However, later on, the contractor was not necessarily the manager. For example, 260 million people have now left the countryside, and even after leaving they still have the contracting right, though they no longer cultivate the land back home. In other words, the contracting right and the management right are separated. The outcome of the combined rights protection is that when the possessor of the contracting right no longer cultivates a parcel of land, he transfers out the management right, but the farmer continues to worry that once he does so, he will never get it back. This problem already occurred during the second contract (January 1, 1997, to December 31, 2026).

In addition, the economic organization of the collective will sometimes weaken the farmers' contracting rights in order to strengthen their management rights. Therefore, the contracting rights and management rights are in fact separated, and nowadays, many problems are encountered in conjunction with the two rights as the system operates.

Finally, there were also big problems in the empowering of rights to land contracting and management. Rights to land contracting and management now encompassed the rights to land tenure, land use, land revenue, and land transfer, but there is no rights to mortgage, guarantee, and inherit before the Third Plenary Session of the 18th CPC Central Committee (2013). As a result, there was a great deal of dispute over rights to mortgage and guarantee whenever rights to land contracting and management were empowered. Everyone worried that when a contractor incurred risk in a mortgage, it would lead to the loss of contracting rights. Therefore, in 2013, the Third Plenary Session of the 18th CPC Central Committee separated the contracting right and the management right: management right holders were empowered to have the right to mortgage, but not the contracting right holders. But did this solve the problem? Not really. Another problem is the lack of inheritance right. Now, among the farming households that had participated in the first round of contracting, the elderly couples that had headed them are long gone. Does their land go back to the collective, based on the contract-issuing and contracting relationship, or does it pass down to the next generation, based on property rights? If it passes to the next generation, is it split among the inheritors, or does it go to the eldest son, as traditionally done in Japan and Europe? The system has no arrangement for this.

3. Choice and Change in the Rural Homestead System

The structure of the rural homestead system is more complicated. Before the reforms, the logic of the rural residential system was not very consistent with the logic of the contracted land system.

First, the private rights of a homestead were always retained by the Advanced Cooperatives—unlike the farmland itself, because at that time it was desired that agricultural land had to serve the purposes of industrialization.

Second, there was the "separation of three rights" (san quan fen li 三权分离): collective ownership of the homestead was separate from the farmer's right to use the homestead, which in turn was separate from private home ownership. That is, ownership belonged to collective, use right was given to the farmers, and the house was private property. Even during the Cultural Revolution, documents put great emphasis on this stipulation: the house was absolutely not transferable. Thus the "separation of three rights" (san quan fen li 三权分离) system was actually formed before the reform.

After this system was formed, the structure of the rural homestead system had the following features:

First, it included a system whereby farmers could obtain their homestead from the collective for free. This was set up after reform, but after the farmers received this parcel of land, what if they wanted to build their own house?

Second, it was impossible in subsequent reforms to strengthen membership status. Before reform, there were many documents that stipulated that once soldiers, teachers, overseas Chinese, and so on left their village (form agricultural household registration to nonagricultural household registration), they would still be allotted homesteads upon their return. After reform, however, when the issue of strengthening membership status was considered, it was decided that only members of the collective economic organization could receive homesteads. In sum, the strengthening of membership rights occurred after the reform, and the establishment of the system whereby farmers could obtain homestead land from the collective for free also occurred after the reform.

Third, the designation of homestead land as a category of land for construction occurred after the reform. No category called "construction land" existed before the *Law of the People's Republic of China on Land Administration* (《中华人民共和国土地管理法》) went into effect in 1987. But what rights appertain to homestead? The *Law of the People's Republic of China on Land Administration* (《中华人民共和国土地管理法》) does not say. After homestead land is classified as construction land used for construction, it is subject to stronger quotas control of land used for construction, and to build a house on it, the farmer must obtain a permit based on the quotas of land used for construction. But it is in fact fundamentally impossible for a farmer building a house to obtain such a quota, because control of the quotas goes down to the county level government, which uses it as a quota for urban construction and economic development. How is it possible that the county government would extend such a quota to a farmer? Therefore, no rights strengthened in the classification of homestead land as construction land used for construction, and only the regulation of that land is strengthened. If the farmer uses it, he gets investigated for breaking the law.

Fourth, government regulation increased. Rights to homestead is a special kind of usufruct. Generally, usufruct includes the right to tenure, use, and revenue, but only the rights to land tenure and land use appertain to the homestead—there is no right to land revenue. Therefore, the right to homestead has become a right of residence that a member obtains free by virtue of his or her status from the collective economic organization.

What have been the consequences of the reform logic of "separation of three rights" (san quan fen li 三权分离) in the homestead system?

First, the fact that the collective locks up the boundaries of each homestead has proven to be more harmful than good. One basic boundary of a homestead is that only a member of the collective economic organization can obtain the homestead, and once that boundary is locked, nonmembers cannot obtain rights to use homesteads. But is this lock effective? Although the boundary is now legally locked, the boundaries of the collective of which the homestead is restricted access are constantly being broken. Along the coast and in suburban areas especially, the divergence between the law limiting homesteads to members of the collective economic organization and the real-world situation grows ever larger.

Second, in comparison with contracted land, collective ownership of homesteads has been realized, but the right of use has been weakened. This is a very big problem that future reform of the homestead system will be facing. In which direction will it be changed? Will collective ownership be further solidified, or will it make right of use even stronger, as it is with contracted land? The difference is now quite large: some people say the former should be stronger, and others lay the latter.

Third, the right to use a homestead and the ownership of a house are now separately empowered. The right to use a homestead is subordinate to the collective—it is the use right under collective ownership—but a house is private. In the documents from the Third Plenary Session of the 18th CPC Central Committee (2013), the farmers' housing property rights may be, as a pilot, mortgaged, used as a guarantee, and transferred, but they do not mention the right to use a homestead—this is because they were thinking of making house ownership more and more important in the future. However, this ownership of the house rests upon the use right of the homestead, and there is a physical indivisibility and division of powers between the two. What should be done? Nothing can be done. The farmers have come up with many ways. In some places, the farmers now basically unite the use rights and their ownership of houses.

Fourth, when a homestead becomes a kind of construction land used for construction, the only obligatory controls on it are the control of quotas and approvals, but no rights empowered to this construction land. The rights that appertain to a homestead include only the right to occupy and the right to use, but not the right to derive income. The difference between the rights appertaining to homesteads and construction land is growing larger and larger. In fact, a great quantity of farmers' homesteads is already being traded. Except for the coastal areas, this is happening basically wherever rural populations are highly mobile.

4. Choice and Change in the Rural Land Conversion System

Let us look at system arrangements. First, the state ownership of urban land and collective ownership of rural land as stipulated in the *Constitution* (1982) amount to the biggest set of systemic arrangements for the conversion of rural land. At first, the design of the two coexisting systems existed for the preservation of the rights of the collective, but after China entered its period of high-speed urbanization, the two ownership systems instead served the process of nationalizing of land.

Second, all converted agricultural land must be expropriated, and it is expropriated according to the land's original use.

Third, the designers of the institutional arrangements claimed to have made reference to modern Western countries' land zoning regulations, but in fact, it was the "three layers of regulation": urban planning, land use regulation, and ownership. Now conversion of land is all a part of urban planning that expands outward from the center. Urban planning is a kind of "hard plan," with hard constraints, and urban governments use urban planning to mark off areas of land. If that is the case, then what use is the land use regulation? If it governs the conversion of agricultural land to construction land, it is a control on the scale of land conversion and on land use. Given that urban planning is a hard constraint, land use regulation is actually a soft constraint.

Why do the cities continue to expand outward? Because ownership regulations stipulate that converting farmland to construction land used for construction can be accomplished only by expropriation based on its original use. Therefore, through planning and regulatory control, the local government finagles it so that land of the farmer's collective finally ends up in its own hands. In this way, the ownership regulations become a tool for the government to obtain land revenue.

Fourth, city and county governments monopolize the primary market of land, and after land is expropriated, they become the exclusive sellers.

The fifth set of systemic arrangements relates to land capitalization and land mortgage financing. The advantage of these arrangements supporting the entire land conversion process is "speed." This is how the faces of entire cities have been able to change all at once, why industrialization and development have grown rapidly, and why China has become the "world's factory" and the world's second largest economy. However, these arrangements have had the following aftereffects.

The first aftereffect is that an entire economy operates along a vicious circle: land sales → local government credit guarantees → land mortgage and financing platforms → reliance on real estate as an export. These four links are now in jeopardy, with land sales, local debt, land mortgages and financing platforms, and real estate all at risk. First, the era of low-cost urbanization, which was originally supported by land sales, has ended, as the cost of land acquisition and demolition has become very high, and local governments' net revenue from land sales has been steadily declining. Then local debt is basically supported by land sales and by taking out new loans to cover old ones. Then there are the land mortgage and financing platforms. Now, finally, real estate is facing a big problem: with land prices and real estate prices so sky-high, who will be able to move into those places?

The second aftereffect is that everyone now advocates "the new urbanization": efficient, inclusive, and sustainable—meaning that the old urbanization has been inefficient, not inclusive, and unsustainable.

The third is a dilemma in land governance. The Third Plenary Session of the 18th CPC Central Committee (2013) proposed the modernization of the country's system and capacity for governance. If the land governance problem is not resolved, this modernization will be difficult to accomplish. The biggest problem here is conflict between officials and ordinary people. The model in which development is sustained through the expropriation of land and through demolition leads to conflict between the government and the people.

The fourth relates to the issue of "new farmers" (*xin nong min*新农民).[3] Formerly it was always said that farmers' issues were always problems of farmers in rural villages. One problem with new farmers now is that they cannot obtain any share of the interests on land—it is a problem of value-added income distribution. Another problem is the dilemma—a very serious one—that exists between farmers who move into a city and the land they leave behind,[4] and it is necessary to find a solution.

5. Land Reform Must Be Done from the Roots

In general, the most important theoretical bases supporting this system are: (1) public ownership; (2) the special nature of land; (3) the land use regulation system; and (4) land value increases return to the public sector.

It cannot be claimed that we have not worked hard to proceed from the "separation of the two rights" (liang quan fen li 两权分离) and the system of immovable ownership to the arrangements for contracted land, homesteads, and the conversion of agricultural land. However, it is obvious that the system has not performed very satisfactorily. If we continue following this reform logic, we will face very big dilemmas, and we need enough wisdom to find a new reform logic.

In my opinion, it is time to reflect on the logic of the separation of the two rights and begin to operate on its roots. The following questions are involved:

One is the reform of collective ownership. If reform of collective ownership is not pushed forward, it will be very difficult to continue using the model of the "separation of the two rights" (liang quan fen li 两权分离) reform. If reform is needed now, which route shall we follow? Under the public ownership system, will it be feasible to change the system of collective ownership?

Let us first consider farmland. Another important institutional design besides "over a long term without change"(*chang jiu bu bian* 长久不变) would be to reform the contract management system so that those who have contracting rights to the land have the rights to land tenure, land use, land revenue, and land transfer. It would make the contracting right real (including the rights to land use, land revenue, and land transfer), and then, after the contracting right is transferred to people who will be cultivating that land in the future, they will have the right of managing it—to mortgage and guarantee.

The second is rural homesteads. The area of homestead reform is the most controversial because everyone worries that farmers will simply sell off their homesteads, causing them to be displaced en masse, so now the talk about homesteads has been on "ensure rural households' usufruct of their homestead, and improve the rural homestead system through reform," and then, "select several pilot areas to steadily and prudently push forward the mortgage, guarantee and transfer of farmers' residential property rights." Basically, this is still designed according to the "separation of three rights" (*sanquan fenli* 三权分离) system.

What is the core of the institutional reform of the homestead? If the usufruct of the homestead are still limited to only the right of tenure and use, that would not amount to any reform. The important issue is to which extent that usufructuary right on the homestead should be expanded.

My suggestion is to expand the usufruct of the homestead, at the very least, to be equivalent to those of usufruct as they are generally understood—namely, the right to tenure, use, and revenue—and then to reform usufruct of homestead by category. For example, in urban-rural integration areas and coastal areas, which no longer distribute homesteads for free, pilot reforms can be carried out on homesteads based on the property rights system. Tenure, use, revenue, and transfer will be treated according to property rights. Reform in more traditional areas can be reasonably postponed.

At the core of another reform for improving the homestead system is the question of how much the system can be reformed distributing homesteads for free according to membership rights.

How can it be improved? My suggestions are, first, to change "free distribution" to "compensated distribution" and to reform the system whereby members are entitled to obtain homesteads. I have already mentioned that in the coastal areas and urbanized areas, there is no more free distribution of homesteads. It may be possible to follow the process of urbanization in these areas and open up the present system whereby boundaries are demarcated according to collective membership status, so that homesteads can enter the land market.

My second suggestion is to reform the different governance of the two ownership systems. If the division of state ownership of urban land and collective ownership of rural land is not reformed, then it will be impossible to make progress in the reform of the system for converting agricultural land. The greatest problem now is how to solve the problem of state ownership of urban land. The most contentious area is the space between built-up urban area and urban planning area. Much of land in this area has not yet been expropriated by the government, but the planning controls do not permit people to use it. Will this collective-owned land continue to be expropriated in the future, or will future reform follow the route suggested by the Third Plenary Session of the 18th Central Committee, of having "marketization of collectively owned rural profit-oriented construction land" (*jiti jianshe yongdi rushi* 集体建设用地入市)? This issue will be at the core of future reform.

Third would be the reform of the land governance system. At its heart, this deals with the relationship between the government and ordinary people, with the core issue being the distribution of land value-added revenue. If the design of the principles of distribution (farmers' shares, pattern of revenue distribution for different classification of farmers, etc.) is not done well, it will become a powder keg in the future.

Finally, we must bring the land issue back to normality; land should no longer drive development but go back to being a condition for development. Recovering normality means changing the model in which local governments must constantly sell land and making them go back to being service providers instead of land business operators. Only on this basis will it be possible to design the arrangements of the entire land system according to modern principles.

If land continues to be an engine for promoting development, the reform of collective ownership, the reform of division of urban and rural land, and the reform of the land expropriation system cannot really progress. Only when normality is recovered on the land issue can other land system reforms be promoted, modern land systems constructed, and society normalized.

Notes

1 "Decision of the Central Committee of the Communist Party of China on Some Major Issues Concerning Comprehensively Deepening the Reform."
2 Separation of ownership rights and use rights.
3 It refers to the outsiders who do not belong to the village but come to live in the village for a long time or engage in agricultural management and other activities. Such outsiders generally have nonagricultural household registration and are not members of the collective economic organization, but their economic activities actually benefit the village.
4 It refers to that some farmers enter the city to obtain nonagricultural household registration or long-term residence. Their economic activities are all carried out in the city. But the right of contracted land still belongs to them. If it can't be transferred, the land will not be used most efficiently.

1 The Historic Origins of China's Land System

Land is the most important resource for human survival and development. More than 2,000 years ago, Mencius said, "There are three treasures of princes: land, people, and governance." *The Great Learning* (《大学》) states: "Once [the ruler] has people, he has land; once he has land, he has wealth; once he has wealth, he has resources for expenditure." More than 300 years ago, William Petty (1623–1687), called the father of British political economy, also famously asserted that "land is the mother of wealth."

A country's land system is its most fundamental system and the basis for the formation of production relations and all economic relations. A good land system has the dual functions of promoting stability and promoting development. By maintaining fairness in land possession and distribution, political stability and social harmony are realized. By protecting land property rights, economic development is promoted. Since the beginnings of human society, in every country, the establishment and improvement of a land system that is compatible with the development of productive forces have become an important part of political and economic institutional changes, and such a land system promotes the evolution and development of society.

1. Historical Evolution of China's Land System

1.1 Basic Context of the Historical Evolution

China's land system, compared with those of other countries, is unique for its long history. When our ancestors were still living in primitive clan communes, they knew how to use land to grow grains and meet the needs for food and clothing. As their civilization developed and the state emerged, the system of public clan ownership evolved into a system of state ownership, as expressed in the poem "The Northern Hills" in the "Minor Odes" section of the *Book of Songs* (《诗经·小雅·北山》): "Under the wide sky, / Everywhere is the king's land. / Within the land's sea-boundaries, / All are the king's subjects." This was the essential feature of the land system during the Zhou dynasty. During the Western Zhou era, China had a "well-field" land system（*jing tian zhi*井田制）, wherein the land belonged to the Son of Heaven, and the various feudal lords and officials had fiefs. This is what was meant by "Everything is the king's land." During the Warring States period, when Shang Yang 商鞅 (c. 390–338 BCE) changed the law, abolished the well-field system, loosened field boundary restrictions, and permitted private ownership of land and the free buying and selling of land, China became the first country in the world to have a private land system. From the Qin and Han dynasties to the middle of the Tang dynasty, China had two parallel land systems, the equal-field (*jun tian* 均田) system of the state and the land occupation (*zhan tian*占田) system of aristocrats, that is, state ownership and private ownership of land. During the middle of the Western Han dynasty, the court adopted a proposal by Dong Zhongshu 董仲舒 (179–104 BCE) and limited the number of acres that aristocrats or nobles could own, depending on their rank, in order to prevent them from annexing land. During the Northern Wei and early Tang dynasties, the equal-field system was promoted. Farming households would divide 100 *mu* (亩) of land depending on how any male laborers they had, ensuring a more even distribution of land. After the "dual tax laws" of the mid-Tang were implemented, China's land system entered a new stage. The dynasties after this no longer implemented the equal-field system, state ownership of land went into decline, and private ownership of land became dominant. By the Song dynasty, China had entered an era of complete private landownership. After the reforms of the "Single Whip Law" (*yi tiao bian fa*一条鞭法; in 1580) and the fusion of the poll tax and the field tax (*tan ding ru mu*摊丁入亩; in 1723–24), the taxation system had become purely a land-based tax, and the state could only rely on the taxation system to adjust the relationship between farmers and landlords and between the state and landlords. The land market was highly developed, and tenancy relationships steadily developed, thus laying the foundation for the development of modern capitalist factors.[1]

After the revolution of 1911, well-intentioned people used the land issue as the key for launching China's modernization, but the results did not meet expectations. In the late nineteenth and early twentieth centuries, Sun Yat-sen, the leader of the Chinese democratic revolution, proposed the "Three Principles of the People" (*san min zhu yi* 三民主义) as an outline for building the nation. At the core of one of those principles, the people's livelihood (*min sheng zhu yi*民生主义) was the idea that "land is to be state-owned, and farmers are to have their fields and fair and equal land rights."[2] After the Kuomintang established the National Government in 1927, it also promulgated the *Land Act* (《土地法》) and the *Enforcement Regulations of the Land Act* (《土地法施行法》) and began carrying out land reform. However, by 1949, when it was defeated and withdrew to Taiwan, the Kuomintang had never managed to achieve "fair and equal land rights," and in fact, the land issue had become more severe.[3]

From the very beginning, the CPC had regarded solving the farmers' land problem as a central issue of the Chinese revolution. After the failure of the National Revolution (of China, 1911), the CPC established rural revolution bases where it carried out an agrarian revolutionary movement, the main purpose of which was to eliminate landownership by feudal landlords. During this agrarian revolution period, the idea of "relying on poor peasants, uniting with middle peasants, restricting rich peasants, protecting small- and medium-sized industrial and commercial workers, eliminating the landlord class, and changing feudal and semi-feudal land ownership into peasant land ownership" was advocated. During the War of Resistance against Japan, there was the proposal to have "landlords reduce rents and interest rates, and farmers pay rent and interest," in order to consolidate the national anti-Japanese united front. During the War of Liberation, when the *Outline Land Law of China* (《中国土地法大纲》) went into force, land owned by landlords was confiscated, the land system of feudal exploitation was abolished, and the policy of every farmer who had his land was put into effect.[4] The success of the CPC on the land issue became the magic weapon in its gain of political power. It was just as Chairman Mao Zedong said as he answered questions from the American journalist Edgar Snow in Yanan in 1936: "Whoever wins over the peasants will win over China, and whoever solves the land issue will win over the peasants."

1.2 Changes in the Land Administration System since the Founding of the New China

Since the People's Republic of China (PRC) was established in October 1949, the CPC has always considered land issues important for stability and development. In the past sixty-plus years, China's land *administration* system has evolved in a manner consistent with the changes in the country's economic and social development. It has been continuously improved through exploration and practice, constantly developed through reform and innovation, and gradually, it has formed a modern land *administration* framework that takes into account China's national conditions and the achievements of civilization elsewhere in the world. The development of China's modern land administration system may be divided into the following periods:

1. *The period of national economic recovery from 1949 to 1952.* During this period, China promulgated important laws of a constitutional nature such as The *Common Program of the Chinese People's Political Consultative Conference* (《中国人民政治协商会议共同纲领》) and the *Land Reform Law* (《土地改革法》; both 1950). At the heart of the land *administration* system was the need to solve the problem of landownership.
   Through the implementation of these two laws, the state-owned land of the old China continued to be accepted, while the land owned by bureaucrats and members of the comprador and bourgeoisie classes was confiscated. In only three years, the largest land reform in modern history, in terms of the number of people affected, was completed: private ownership of landlords was abolished, and farmers' landownership was established. More than 300 million farmers who have no or little land have obtained more than 700 million acres of land and other means of production, which has greatly mobilized farmers' enthusiasm for production. Rural productive forces and farmers' living standards have improved significantly.[5]
2. *The period of socialist transformation and exploration from 1953 to 1977.* During this period, China promulgated the first *Constitution of the People's Republic of China* (《中华人民共和国宪法》; 1954), formulated the *Draft Model Charter of Agricultural Production Cooperatives* (《农业生产合作社示范章程草案》; 1955), *Model Regulations for Advanced Agricultural Producers' Cooperatives* (《高级农业生产合作社示范章

程》; 1956), and the *Draft Amendment of the Regulations Governing Rural People's Communes* (《农村人民公社工作条例修正草案》; 1962; called the "Sixty Articles on People's Communes"), and came out with the *Regulations on the Requisition of Land by the State for Construction* (《国家建设征用土地办法》; promulgated 1953, amended 1958). The focus of the land *administration* system was to establish rural collective landownership as well as a national system for the requisition of land for construction purposes.

In the countryside, the 1954 *Constitution* clarified that the state, in accordance with the law, protects farmers' ownership of land and other means of production. Soon afterward, through the agricultural cooperative movement of farmers' mutual aid organizations—the elementary communes (初级社)—and advanced communes (高级社), the socialist transformation of rural areas was basically completed, and the transformation from rural private landownership to public landownership was realized. In 1958, the People's Commune Movement, summed up in the slogan "One Big, Two Public, and Three Leveling (*yi da, er gong, san la ping*一大、二公、三拉平),"[6] was launched in order to make advances in the areas of landownership system, rural production, and rural lifestyle. However, after three years of natural disasters, there was a return to the rural collective ownership system characterized by the slogan "Three-level Ownership, with the [Production] Team as Basis (*san ji suo you, dui wei ji chu*三级所有、队为基础),"[7] and the "Sixty Articles on People's Communes" provided the legal basis for its establishment.[8] At the same time, China launched economic reconstruction on a large scale, and as industrialization was promoted throughout the country, the issue of state land expropriation for construction use became prominent. The *Regulations on the Requisition of Land by the State for Construction*, which was promulgated by the State Council, clarified the purposes of the state's requisition of farmers' land and the principles of compensation and resettlement. For the first time, the statement, "For the needs of the public interest, the state may, in accordance with law, engage in the acquisition, expropriation, or nationalization of urban and rural land," was written into the Constitution.

3. *The period of reform and opening up from 1978 to the present.* During this period, China amended the *Constitution of the People's Republic of China* three times, promulgated its first *Law of the People's Republic of China on Land Administration* (《中华人民共和国土地管理法》; later

thoroughly revised), and put out a series of laws and regulations such as the *Provisional Regulations of the People's Republic of China on the Transfer and Transfer of the Right to Use of State-owned Land* (《中华人民共和国城镇国有土地使用权出让和转让暂行条例》) and the *Law of the People's Republic of China on Urban Real Estate Administration* (《中国人民共和国城市房地产管理法》). It enacted the *Real Right Law of the People's Republic of China* (《中华人民共和国物权法》), the *Law of the People's Republic of China on Rural Land Contracting* (《中华人民共和国土地承包法》), and so on. China's land *administration* was thus put on an institutionalized and legalized track, focused on the formation and gradual improvement of China's modern land *administration* system.

Between 1978 and 1986, China formed several major land *administration* systems. In rural areas, the people's commune system of "political and social unity (*zheng she he yi*政社合一)" was abolished, and in its place the household contracting management system under the collective ownership system was implemented. The 1982 revised *Constitution* for the first time clarified the legal framework for the coexistence of the two ownership systems, wherein "urban land is state-owned and rural land is under the collective ownership of the farmers." The first *Law of the People's Republic of China on Land Administration*, promulgated in 1986, further clarified the basic framework of landownership:

Land in urban areas belongs to the whole country and is owned by the state. Land in rural areas and suburban areas of cities belongs to collective ownership except where law stipulates that it belongs to the state; homesteads, private reserved plots, and private reserved mountains[9] belong to collective ownership.

It stipulated methods for land users to obtain and use construction land and rural land under state ownership, and it clarified the principles of centralized, uniform management of land throughout the country and the use of land use master plans for the management of land used for construction.

From 1987 to 1997, the focus of the land *administration* system was establishing a system for the paid sale of state-owned land-use rights in urban areas that was compatible with the socialist market economy. For a long time during the planned economy period, China had implemented a system in which state-owned land was used "for free, for an indefinite period, and with no transfer permitted." Successful trial runs of the sale of state-owned land-use rights in cities such as Shenzhen provided an opportunity for promoting the system of paid use of urban state-owned land. In 1988, the *Constitution* and *Law of Land Administration* were amended to remove stipulations against the leasing out of land, and it was clarified that "land-use rights may be transferred in accordance with the provisions of the law." Following the reforms of the land contracting and use systems under the collective ownership system in rural areas, a system separating ownership and use rights was also implemented for China's urban state-owned land. The *Provisional Regulations on the Transfer and Transfer of the Right to Use of State-owned Land* and the revised *Law of Urban Real Estate Administration* and *Law of Land Administration*, promulgated in 1990 and 1994, respectively, marked the formal establishment of the market trading system for the sale, transfer, lease, and mortgage of urban state-owned land-use rights in China.[10] The reform of the system of paid use of state-owned land in cities and towns has played an important role in realizing the rights of landowners, demonstrating the value of state-owned land assets, and promoting urban construction.

Since 1998, China's land *administration* system has entered a period of further improvement. The 1998 revision of the *Law of Land Administration* has led to major reforms in methods of land *administration* and use. Based on China's basic national condition of high population density relative to land, and drawing from the experiences of land *administration* in advanced countries, the basic framework for a modern land *administration* system was defined—a system which aims to protect cultivated land, uses regulations on use as the means, and places overall land use planning in front. It established a basic system for protecting cultivated land, a system for offsetting cultivated land used for construction purposes (in order to keep the amount of cultivated land unchanged or not reduce), and a system for approving applications to convert farmland to construction land used for construction—all for strictly controlling such land conversions and effectively protecting cultivated land. Coming into the 2000s, the CPC Central Committee successively issued relevant documents on new situations or problems arising in land *administration*, such as:

"Notice on Further Strengthening Land *Administration* to Effectively Protect Cultivated Land 《关于进一步加强土地管理切实保护耕地的通知》" (CPC Central Committee and State Council [中发], 1997, No. 11);

"Decision on Deepening Reform on Strict Land *Administration* 《关于深化改革严格土地管理的决定》" (CPC Central Committee and State Council, 2004, No. 28);

"Notice on Issues Pertaining to the Strengthening of Land Control 《关于加强土地调控有关问题的通知》" (State Council Decree [国发], 2006, No. 31);

"Notice on Issues Pertaining to the Establishment of a National Land Monitoring System 《关于建立国家土地督察制度有关问题的通知》" (State Council General Office [国办发], 2006, No. 50);

"Notice on Promoting Economical and Intensive Use of Land 《关于促进节约集约用地的通知》" (State Council Decree, 2008, No. 3);

"Notice on Strict Implementation of Laws and Policies on Rural Collective Construction Land Use 《关于严格执行有关农村集体建设用地法律和政策的通知》" (State Council General Office, 2007, No. 71).

These strengthened strict management according to law, initiated the inclusion of land in macroeconomic regulation and control, established a national land monitoring system, and proposed the implementation of "the two most stringent" land systems (i.e., the most strict systems for protecting arable land and ensuring the most economical and intensive use of land), and thus enabled China's land *administration* system to improve steadily.[11]

During this period, historic progress was made also in the establishment of China's land rights laws. The *Law of the People's Republic of China on Rural Land Contracting* (《中华人民共和国农村土地承包法》), promulgated in 2002, marked the realization of legal protection for rural land rights. The *Real Right Law of the People's Republic of China* （*《中华人民共和国物权法》*） of 2007 marked the official establishment of China's system for the protection of property, primarily real estate such as land, so that land rights protections were put on a legal track.

In short, after more than 60 years of hard work, China has established a complete legal system for land consisting of laws and regulations in the areas of constitution, Real Right Law, civil law, land *administration* law, forestry and grassland law, environmental protection law, and so on. The framework and basis for a modern land *administration* system that befits Chinese socialism in the initial stage and the socialist market economy with Chinese characteristics have been formed. The various basic established systems are being constantly enriched and perfected through practice, and innovations are made in the forms in which they are implemented.

2. The Main Features of China's Modern Land *Administration* System

2.1 A Land Rights System Based on the Socialist Public Ownership of Land, the Coexistence of Two Ownership Systems, and the Separation of Ownership and Use Rights

The land system is an important part of any country's basic economic system, and its institutional characteristics are regulated by a state's basic system. For example, in the United States, which is a country that has private ownership and a federal system, 58% of the land is privately owned, while the vast majority of the forests, grasslands, protected areas, and marshes are owned by the federal and state governments (31% and 10%, respectively). City governments own approximately 1% of the land.[12] By contrast, according to China's *Constitution*, the socialist system is China's fundamental system, and socialist public ownership of the means of production is the foundation of China's socialist economic system. Insisting on "maintaining the dominant role of public ownership while allowing for different economic activities in diverse forms of ownership" is the country's basic economic system at the initial stage of socialism. As an important component of the country's basic economic system, the structure of China's land system is manifested in the following ways:

(1) *It implements the socialist public ownership of land.* No organization or individual may misappropriate, buy, sell, or otherwise illegally transfer land.
(2) *State ownership of land coexists with collective ownership of land.* Mineral deposits, waterways, forests, mountains, grasslands, wastelands, beaches, and tidal flats, and other natural resources are state-owned, and urban land is also state-owned. The land in rural areas and city suburbs is collectively owned, except where law stipulates that it is state-owned. Rural residential land (homesteads), private reserved land, and private reserved mountains are collectively owned; where law stipulates that they are collectively owned, forests, mountains, grasslands, wastelands, beaches, and tidal flats belong to collective ownership.
(3) *Landownership is separate from land use rights.* A rural land contract management system has been adopted for rural land (which refers to cultivated land, forests, grasslands, and other land used for agriculture by

farmers' collectives in accordance with law, whether collectively owned or state-owned), and this gives farmers long-term and guaranteed land use rights. For state-owned land, a system of paid use has been implemented; under the premise of state ownership, land use rights are sold to land users for a specified period of time.

(4) *Land property rights are respected and protected.* The land rights system is at the core of the land system. The legal protection of land rights can enhance the long-term expectations of rights holders to use land, and it promotes economic development and the optimal distribution of resources. For a long time, due to the constraints of the planned economy system and traditional ideology, China did not adequately protect property rights. During the planned economy period, it did not even recognize individual property rights, which affected the healthy development of the social economy. After reform and opening up, as institutional reforms deepened and the scale of market allocation of resources increased, the function of property rights in economic development kept growing, and people became increasingly aware of their rights. The improvement of the socialist market economic system also called for rights guarantees and protections. China made great progress in its land rights legislation. Now a legal system that regulates land rights by the *Constitution*, the *General Principles of the Civil Law* (《民法通则》), the *Real Right Law*, the *Rural Land Contract Law*, and the *Land Administration Law* has been established. Provided that the ownership of the land remains unchanged, the strengthened and guaranteed usufruct (such as the right to contract and management land, use right of homestead, and use right for construction land used for construction) protect the rights of land users.

2.2 Protecting Cultivated Land: The Primary Goal of China's Land Administration System

Due to its high population density relative to land, shortage of high-quality arable land, and mismatch between water and soil resources, China attaches primary importance to food security and protecting arable land for its economic development, during the present stage of accelerated industrialization and urbanization. The *Law of Land Administration* of 1998 clarified that the goal of land administration is to "strengthen land *administration*, safeguard the socialist public ownership of land, protect and develop land resources and the rational use of land, effectively protect cultivated land, and promote the sustainable development of the social economy." It proposed that the "very cherished and rational use of land and effective protection of cultivated land are the basic national policies of our country." The world's strictest system for protecting cultivated land is now being implemented through the arrangements of an entire series of systems, such as providing special protection to permanent basic cropland, offsetting cultivated land used for other purposes, intensifying efforts to develop, improve, and reclaim rural land, and responsibility system for protecting cultivated land. In 2008, a red-line target of no less than 1.8 billion *mu* (equivalent to 124 million hectares) in the country before 2020 was proposed.

2.3 Establishing a Land Administration System with Land Use Regulations as the Core

Land use regulation is a system adopted by countries with relatively complete land use systems. Japan, the United States, Canada, and other countries call it "zoning control"; in Sweden, it is called "land use regulation"; in the United Kingdom the "land planning permit system"; and in France and South Korea the "construction and development permit system."[13] In the perspective of these countries, land use regulation has achieved good results in the rational use of land resources, especially in the protection of cultivated land. During the planned economy period and the beginning of the reform and opening-up period, land administration in China depended mainly on ownership regulation, and the state would either transfer ownership without compensation or requisition farmland with very low compensation and convert it to state-owned land. When the *Law of Land Administration* was amended in 1998, China brought in Western regulation systems of proven effectiveness and wrote them into law. China's land use regulation system is thus a system in which land use master plans are prepared, and these demarcate zones of specific land use and determine land use restrictions. Landowners and land users then use the land in strict accordance with these plans.

China's land administration system with land use regulations includes four main parts. First, the state prepares the overall plan of land use to stipulate land use, and divide land into farmland, land used for construction, and unused land. Second, the law stipulates the status, function, and approval procedures of the overall plan of land use. Third, the state strictly limits the conversion of agricultural land into construction land, controls the total amount of construction land, and implements special protections for cultivated land. Fourth, the law clearly stipulates the approval authority for the conversion of agricultural land into construction land and appropriately centralizes the approval authority for land expropriation.

2.4 Implementing a Mainly Centralized and Uniform Land Administration System

Generally speaking, the world's land administration systems have evolved through a process away from divided, decentralized, multi-administration to centralized administration, and the same is true for China's land administration system. Before 1986, China's land administration was dispersed among many administrative authorities, subject to multi-administration. After the market economy was implemented, the drawbacks of the "integration of departments and regions at different levels, taking regions as the core" land administration system became increasingly obvious, and they resulted in a loss of control of land and a sharp decline in cultivated land. In 1986, due to its special national conditions and scarce resources, China established a unified national urban and rural land administration system:

The land administration department of the State Council[14] shall be in charge of the unified administration of national land. The land administration departments of local people's governments at or above the county level shall be in charge of the unified administration of land within their respective administrative areas. Township-level people's governments shall be responsible for land management within their respective administrative areas.[15]

After 1998, the unified administration of land across the country was extended to include minerals and marine resources. In 2006, the State Land Supervision System (国家土地督察制度) was established. By now, China has formed a relatively centralized, top-down supervisory, unified land management model. However, at present, agricultural land, forest land, and grassland are still separately managed by relevant departments in accordance with relevant laws.

3. Important Enlightenment from History

From the evolution of China's land system through history, we find the following enlightenment:

3.1 Land Issues Are Related to National Politics and Social Stability

Land issues have always been closely related to China's order and the people's well-being; they have intimately connected with social stability and the succession of dynasties. The land system, with the distribution of land at its core, often plays a vital role in a country's political and social stability. Judging from historical records, rulers throughout history were all concerned about small farmers' land issues. They saw that as long as the peasants had food, society would be stable.

For more than 2,000 years, through the development of China's feudal society, the state has attached great importance to the use of land policies for maintaining political stability. During the Warring States era, the state of Qin adopted Shang Yang's（商鞅） ideas and promoted the land allotment system (*shou tian zhi* 授田制), which featured the "opening of field boundaries" (kai qian mo 开阡陌), enabled the rapid development of agricultural production, and laid the foundation for Qin's rise and the unification of the six remaining states. In the early days of the Tang dynasty, the government vigorously promoted the equal field system (均田制) in order to restore agricultural productivity, which had been severely damaged by tyranny and warfare since the end of the Sui dynasty, and this brought economic recovery and prosperity. In a short span of more than 100 years, from the Sui through the early Tang, there were several periods of peace and prosperity (rare in Chinese history), such as the Kaihuang period (581–600) under the Sui Emperor Yang, the Zhenguan period (627–649) under the Tang Emperor Taizong, and the Kaiyuan period (713–741) under the Tang Emperor Xuanzong, and in each case it was because the feudal government in those periods came out with land policies that suppressed land consolidation, reduced taxes, and allowed for rest and recuperation.

In China's history, there were many other periods in which the neglect of land issues or faulty land policies had disastrous results. During the mid- to late periods of most Chinese dynasties through the ages, there were rather large peasant uprisings, and nearly all of these peasant armies were made up of landless peasants. Furthermore, most of them had been displaced precisely because of land issues. When land consolidation caused an increase in the number of displaced people (who lived below the line of viability), then a one-time event (like a mutiny, an external disaster, natural disaster, etc.) would be enough to cause these displaced people to join a rebellion. These rebellions would then function like "the last straw," breaking the back of the dynasty.[16]

Again, after the founding of the New China, we have had land issue successes and failures from which we can learn. The land reforms that took place in the early days of the PRC and the reforms of the late 1970s, when the rural land contract responsibility system was put in place, released pent-up agricultural productivity, but the people's commune movement of the late 1950s caused the destruction of agricultural productivity to a considerable extent.[17]

In short, land issues are related to a country's political and social stability. Because the land system plays an important role in promoting or restricting social stability and development, one must attach a high degree of importance to it.

3.2 Dealing with the Distribution of Land Interests Is a Major Concern for Social Stability and Development

Land issues are extremely complicated, but one of their major components is the problem of the distribution of land benefits. It was one of the major factors in the cycle of dynastic replacement and succession mentioned above.

From the historical point of view, the distribution of land benefits consisted of two kinds of distribution: (1) the distribution of land itself, that is, land rights; and (2) the distribution of agricultural produce from the land, that is, land rent. Basically, on the one hand, the rule of the state depended on support from powerful landlords, though there was also some conflict between landlords' interests and those of the state. On the other hand, the expenditures of the state and the landlords depended on payments in tax and rent from the peasants, so that political instability could also be caused when peasants' interests exceeded what they could bear in rent. Historically, privileges of powerful or noble landlords were stipulated by law. However, the wealthy would generally conceal a large amount of their real estate and peasant population and keep them private, which resulted in reductions in state tax revenue and loss of state interests. Not only that, landlords would often use their strong positions to deprive peasants of their interests; they would consolidate their land holdings and pass the burden on to the peasants. When the peasants were no longer able to bear the costs, they become displaced people, which in turn increased the burden on the state to maintain order.

Therefore, all of the more capable rulers in Chinese history were good at handling the distribution of land interests. While they suppressed powerful landlords and kept them under control, they would ease conditions for the peasants and thereby achieve a balance of interests. When the Cao Wei dynasty (220–266) implemented its famous *tuntian* (屯田) system, the advice given to the dynastic founder Cao Cao 曹操 (c. 155–220) was that he "should follow their [the peasants'] desires (宜顺其意)."[18] When the court of the Northern Wei dynasty (386–534) was discussing the equal field system, the advice given was "follow the people's desires (因民之欲)."[19] One of the most ancient and most exquisite statements of political philosophy preserved in Chinese literature asserts: "The people are the roots of the country. If the roots are secure, the country is at peace (民惟邦本，本固邦宁),"[20] and it is this spirit that runs through the land system and its laws.[21]

For more than 60 years in the New China, the government had experiences and lessons in dealing with the interest relations of the state and farmers, and these are worth summing up. In the traditional planned economy period, the cost differential between the unified purchase and sale system and the discounted prices for agricultural products provided for the country's industrialization, but it also led to tension between the state and farmers as well as poverty among the farmers. After 2000, the government abolished the "royal and imperial grain tax " (皇粮国税) that had existed for thousands of years, and the increase in farmers' incomes has promoted the stability of rural society. As industrialization and urbanization accelerate, the increases in land value-added income will bring new issues in the distribution of land interests, which must be given serious consideration.

3.3 The Land Administration System Plays an Important Supporting Role in Economic Development

The Chinese nation is a long-standing, independent agricultural civilization. In the agricultural civilization period, China was among the few most advanced and developed countries in the world, and closely related to the development of this agricultural civilization was China's land management system.

China was the first country in the world to carry out land registration, institute property rights protections, and implement land administration laws. Before 2000 BCE, under the Xia dynasty, land surveys were already carried in each of the Nine Regions (*jiu zhou* 九州). Some inscriptions on bronze vessels from the Western Zhou period (1047–772 BCE) had numerical counts of fields, and during and after the middle of the Spring and Autumn era (771–476 BCE), the states of Lu 鲁, Chu 楚, and Zheng 郑 successively conducted field and land surveys for the purpose of taxation. Starting in the Warring States period (475–221 BCE), private ownership of land was determined by law. Beginning in the Qin (221–206 BCE) and Han dynasties (206 BCE–220 CE), the small-scale peasant economy, with each household the basic unit, was recognized by law, and this gradually became dominant. After the mid-Tang and especially after the Song dynasty, private landownership increasingly flourished, and the cadaster gradually gained a status parallel with household registration. After the "Single Whip Law (*yi tiao bian fa* 一条鞭法)" went into effect in 1580, in the Ming dynasty, the "fish-scale atlases (鱼鳞图册)" became the main basis for corvée labor recruitment and cadaster management.[22] Through the imperial dynasties, the direct purposes of compiling household registers and cadaster were conscription and taxation, and the cadaster had an additional role in protecting land property rights.

Given the historical conditions of those times, this family-based, clearly established ownership land system was conducive for maintaining farmers' enthusiasm and for the optimal distribution of resources. It stood in stark contrast to the dominant economy of Western feudal society. The period from the Song dynasty to the middle of the Qing dynasty was the peak of China's agricultural economic development; China's economic aggregate and living standards were always at the world's forefront. The implementation of the rural household contract responsibility system in the early 1980s opened the curtain on China's reform and opening up, legally established the family-based management system as China's basic management system, and protected farmers' land property rights according to law, and this has played a fundamental role in promoting China's rural economic development and social stability.

3.4 Too Many People, Too Little Land: The Perennial Constraint on China's Land Administration System and Economic Development

Despite China's vast land area, the relative lack of useful land and excessively rapid population growth led to a low amount of arable land per capita, so that there always existed a dilemma between the number of people and the availability of land. The pre-Qin thinker Han Fei Tzu 韩非子 (279–233 BCE) discovered, "Today if a man has five sons, it is not thought too many, and if those sons each have five sons and the grandfather has not yet died, then there are twenty-five grandsons. For this reason, the people are multitudinous but goods and money are scarce; the work becomes arduous but supplies little of sustenance."[23]

China's agricultural civilization has for more than 2,000 years stimulated population growth. During the Spring and Autumn and Warring States eras, the population already exceeded 20 million, and by the Western Han dynasty (206 BCE-9 CE), the population reached 60 million. Starting from the middle of the

Tang dynasty, though cycles of population increase and decrease occurred due to major disturbances or dynastic changes, the overall trend was that China's per capita arable land area declined due to population growth. In the Northern Song dynasty, the population exceeded 100 million; in the Ming, it reached 140 million; and in the Qing, it broke the 400 million mark. After 1741, the sixth year of the Qianlong Emperor's reign in the Qing, China's population continued to grow at high speed, doubling every 25 to 45 years. The population pressure on the land began to get much more serious starting in the early years of the Republic of China era. Because the population in all the core areas of the major river basins has basically reached the saturation point since the Ming and Qing dynasties, the continued population increase has led either to migration to marginal areas or to increasingly labor-intensive farming in the core areas, with diminishing marginal returns. The migration to marginal areas led to an excess of wasteland, due to overexploitation of resources, and ecological problems. For example, during the Northern Song dynasty, excessive land reclamation, excavations in mountainous terrain in order to create new farmland, and extending fields into lakes to create new farmland led to two major breaches of the Yellow River. The river's course changed, which had serious consequences. Meanwhile, the increasingly labor-intensive farming in core areas meant that on family farms, more and more labor had to be invested just to meet survival needs, until the marginal remuneration fell to nearly zero. This caused widespread poverty among farmers.[24]

In addition, due to the long-term implementation through Chinese history of the policy of encouraging agriculture and constraining commerce, China also lost the opportunity to resolve population pressure through modern industrial civilization. In the traditional agrarian perspective, there was no way of solving the problem of the peasants—they could only be caught in a vicious circle between too many people and too little land. Thus in early modern times, China lagged further and further behind the industrializing nations. During the period of collectivization, as the population was tied to the land, China's agricultural involution became even more serious, and farmers fell into poverty.[25]

Since reform and opening up, with the acceleration of industrialization and urbanization, more and more farm laborers are leaving the farms, and fertility rates are changing, too, so that China is encountering a rare opportunity to resolve the problem of stalled modernization due to population pressure. However, accelerated industrialization and urbanization and the movement of large numbers of people away from the farms have also brought about dilemmas relating to resources, and these have become prominent problems facing China as it achieves the goal of building a moderately prosperous society in all respects.

Notes

1 See Fang Xing 方行, "Zhongguo fengjian jingji fazhan jieduan shulüe 中国封建经济发展阶段述略 [The Stages of China's Feudalist Economic Development in Brief]," *Zhongguo jingjishi yanjiu* 《中国经济史研究》[Researches in Chinese Economic History](2000.4): 14–26.

2 See Sun Wen 孙文, *Sanmin zhuyi* 《三民主义》 [*Three Principles of the People*] (Beijing: Jiuzhou Publishing House, 2012).

3 See Jin Dequn, ed., *Zhongguo guomindang tudi zhengce yanjiu* 《中国国民党土地政策研究》 [*Study of Kuomintang Land Policy*] (Beijing: China Ocean Press, 1991).

4 See Zhongguo shehui kexueyuan jingji yanjiusuo xiandai jingjishi zu 中国社会科学院经济研究所现代经济史组 (Modern Economic History Division, Institute of Economics, Chinese Academy of Social Sciences), ed., *Zhongguo tudi gaige shiliao xuanbian* 《中国土地改革史料选编》 [*Selected Historical Materials on China's Land Reform*] (Beijing: National Defense University Press, 1988).

5 See Zhongguo shehui kexueyuan jingji yanjiusuo xiandai jingjishi zu 中国社会科学院经济研究所现代经济史组 (Modern Economic History Division, Institute of Economics, Chinese Academy of Social Sciences), ed., *Zhongguo tudi gaige shiliao xuanbian* 《中国土地改革史料选编》 [*Selected Historical Materials on China's Land Reform*] (Beijing: National Defense University Press, 1988).

6 *Translator's note:* The slogan meant that each commune was to be large in scale (size, membership, and production), that everything was to be publicly owned by the commune, and that differences both within and among communes were to be eliminated.

7 *Translator's note:* That is, the means of production (including land) were owned on three levels, and the production team was taken as the basic accounting unit.

8 See zhonghua renmin gongheguo guojia weiyuanhui bangongting 中华人民共和国国家农业委员会办公厅 (General Office of the State Agricultural Commission of the people's Republic of China), ed., Bongye Jiti Hua Zhongyao Wenjian Huibian 《农业集体化重要文件汇编》 [Compilation of important documents of agricultural collectivization] (Beijing: CPC Central Party School Press, 1981).

9 *Translator's note*: Private reserved plots (*zi liu di*自留地) and private reserved mountains (*zi liu shan*自留山) are lands set aside for farmers to grow their own crops and other private purposes, but they are still under collective ownership.

10 Huang Xiaohu 黄小虎, ed., *Xin shiqi Zhongguo tudi guanli yanjiu* 《新时期中国土地管理研究》 [*Study on Chinese Land Administration in the New Era*] (Beijing: Dangdai Zhongguo chubanshe, 2006).

11 Gan Zangchun 甘藏春, "Woguo tudi guanli tizhi gaige de mubiao, yuanze he neirong 我国土地管理体制改革的目标、原则和内容 [*The Objectives, Principles, and Content of the Reforms to Our Country's Land Management Systems*]," internal report.

12 Kong Qingshan 孔庆山, *Meiguo zaoqi tudi zhidu yanjiu* 《美国早期土地制度研究》 [*Study on Early American Land Systems*] (Guangzhou: Sun Yat-sen University Press, 2002).

13 Liu Xinwei 刘新卫, *Zhuyao guojia guihua he yongtu guanzhi zhidu* 《主要国家规划和用途管制制度》 [*Planning and Use Control Systems in Major Countries*] (Beijing: China Land Press, 2010).

14 *Translator's note*: Over time, this has meant the State Administration of National Land (1986–1998), the Ministry of Land and Resources (1998–2018), and since 2018, the Ministry of Natural Resources.

15 The *law of the people's republic of china on land administration* (Adopted at the 16th Meeting of the Standing Committee of the Sixth National People's Congress on June 25, 1986).

16 Liang Fangzhong 梁方仲, *Zhongguo lidai hukou, tiandi, tianfu tongji* 《中国历代户口、田地、田赋统计》 [*Statistics on Household Registrations, Farmland, and Field Tax Through Chinese History*] (Shanghai: Shanghai People's Press, 1962).

17 Du Runsheng 杜润生, *Du Runsheng wenji* 《杜润生文集》 [*Collected Writings of Du Runji*] (Taiyuan: Shanxi Economic Press, 2008).

18 *Translator's note*: Under the *tuntian* system, landless peasants, refugees, and soldiers were assigned to plots of land which they were to farm, while farming implements were provided by the government at a low price. In exchange, the peasant was to give over half of his harvest to the government. It was very successful due to the large quantity of abandoned farmland from the previous decades of civil war. The statement on following the peasants' desires was part of Yuan Huan's 袁涣 (*fl.* 190–210) advice to Cao Cao; see Chen Shou 陈寿, *San guo zhi* 《三国志》[*History of the Three Kingdoms*] (Beijing: Zhong Hua Book Company, 1960), 11.334.

19 *Translator's note*: The advice was given by Li Zhong 李冲 in 486; see Wei Shou 魏收, *Wei shu* 魏书 [*History of the Northern Wei*] (Beijing: Zhong Hua Book Company, 1974) 53.1180.

20 Translator's note: The ancient statement comes from "Wu zi zhi ge 五子之歌 [Songs of the Five Sons]" in Shang shu 尚书 [The Book of Documents].

21 Fang Xing方行, "Zhongguo fengjian jingji fazhan jieduan shulüe 中国封建经济发展阶段述略 [*Stages in the Development of China's Feudal Economy, in Brief*]," *Zhongguo jingjishi yanjiu* 《中国经济史研究》[Researches in Chinese Economic History] (2000.4): 12–24.

22 See Liang Fangzhong 梁方仲, Zhongguo lidai hukou, tiandi, tianfu tongji 《中国历代户口、田地、田赋统计》 [Statistics on Household Registrations, Farmland, and Field Tax Through Chinese History] (Shanghai: Shanghai People's Press, 1962).

23 Translator's note: The ancient statement comes from "Wu Du 五蠹 [Five Beetles]" in Han Fei Zi Ji Jie 韩非子集解 [Subset Solution of Han Fei Tzu]. (Beijing: Zhong Hua Book Company, 1998), 19. 443.

24 Dwight H. Perkins, *Zhongguo nongye fazhan, 1368–1968* 《中国农业发展，1368–1968》 [*Agricultural Development in China, 1368–1968*] (Shanghai: Shanghai Translation Press, 1981).

25 Huang Zongzhi 黄宗智, *Zhongguo de yinxing nongye geming* 《中国的隐性农业革命》 [*China's Hidden Agricultural Revolution*] (Beijing: China Law Press, 2010).

2 China's Rural Land System: Features and Problems

China's current rural land problems can be broken down into three main areas: farmland contracting management rights, homestead, and rural collective construction land. This chapter will sort out how the corresponding rural land systems were formed, the systems' features, and the problems that have arisen in each of the three areas.

1. The Farmland Contracting Management Rights System: Features and Problems

Debates and experiments over the rural land system did not stop after the CPC gained political power. In the decades since then, policymakers in China have constantly debated the possible choices and changes within this core system. When assessing innovations to rural systems and exploring their future trends, we cannot ignore the current situation in the rural land system and groundlessly debate the issues. This section will briefly review the formation of the rural contract management system and, on this basis, summarize its main features.

1.1 The Lessons and Legacy of Agricultural Collectivization

The land reform undertaken between 1949 and 1951 was the CPC's "first battle of the great, comprehensive revolution upon entering the countryside" after it had obtained political power. It was not merely a change of economic system through land redistribution, abolition of landlord landownership, and reestablishment of peasant landownership. It was also a "dynastic change," a political change. That is, by "mobilizing the masses to reorganize the grassroots," completely overthrowing the old rural order, and establishing the peasant masses' political advantage, the legal foundation was laid for realizing the CPC's political and economic leadership in the countryside.[1]

The CPC's land reforms differed not only from the expedient equal-field relief measures of the past imperial dynasties but also from the land reforms carried out in various countries and regions after World War II, which aimed to solve the problem of land inequality without changing the landownership system. It was the first battle of the revolution to destroy the land property system based on private ownership and the agricultural management system based on the small farmer. At the highest decision-making level, the only differences of opinion were over the length of time for retaining the farmers' private ownership of land and the preparation of conditions for abolishing this system. When the New China was first founded, the people who were implementing rural policy focused on guiding farmers spontaneously and voluntarily, within a certain scope and scale, uniting together to form mutual aid groups based on private ownership, with the farming household as the basic unit. However, not long after this land reform, the buying and selling of land reemerged in rural areas, and differentiation among farm households appeared. To prevent the return of old rural production relations, Shanxi, northeast, Hebei, and other "old revolutionary base areas" changed a few of the mutual aid groups into agricultural cooperatives, with land shares; these implemented land-based dividends and distribution based on work, and they accumulated public funds and public property in an attempt to shake up private ownership. These approaches relating to the direction of the systemic change led to fierce debates in the upper levels of the central government.[2] Liu Shaoqi severely criticized these attempts: "Using cooperatives to shake up private ownership is erroneous, dangerous, and imaginary agricultural socialist ideology." Mao Zedong, on the other hand, expressed clear support: "You mean that we can't organize farmers into communes, rely on the division of labor and teamwork, and unify operations in order to increase productivity and shake the foundations of private ownership?" The difference between Mao and Liu was one of strategies, on how to achieve collectivization: Liu recommended industrialization first, and then agricultural collectivization, while Mao argued that even in the absence of industrialization, collectivization could be carried out first through adjustments in production relations. A bigger difference was that Liu, following the *Common Program of the Chinese People's Political Consultative Conference* (《中国人民政治协商会议共同纲领》), wanted China to carry out a new democratic revolution first for a period of time and then, through the socialist revolution, to abolish private ownership, including farmers' private landownership, while Mao had already started to favor directly entering the socialist revolution.[3]

In 1951, under the leadership of Mao Zedong, the idea of "planning to complete the task of collectivization within 10–20 years and transform the small-scale peasant economy" was put forward. In 1953, the General Economic Guidelines in the Transition Period (1949–1956) in which the country would "basically realize the industrialization of the country as well as the socialist transformation of agriculture, handicrafts, capitalist industry, and commerce" was proposed. The stage during which the new democracy was to be established was deemphasized, and the time for the socialist "general offensive," which was originally planned to be launched 15–20 years later, was moved ahead. Thus, after less than three years of widely appreciated land reform, the farmers were organized and swept into the vast agricultural collectivization movement.

As directed by Mao Zedong, the collectivization movement quickly began. In 1950, there were only 19 agricultural production communes in rural China (including 18 elementary communes and one advanced commune). By the end of 1951, the number had grown to 120 (129 elementary communes and one advanced commune). In December 1953, the CPC Central Committee issued a "Resolution on Agricultural Production Communes" (《关于农业生产合作社决议》) that greatly accelerated collectivization, so that the number of such communes grew to more than 15,000 (including 15,000 elementary communes and 15 advanced ones). In 1954, the number had grown to more than 114,000 (including 114,000 elementary communes and 200 advanced ones). As the number of communes climbed, their development was accompanied by various negative phenomena and coercion: new communes collapsed and their members dispersed, members resigned, large amounts of livestock were sold off and sheep slaughtered, large numbers of trees were cut down, and the situation in the countryside became tense. The Central Rural Work Department, which was responsible for this work, made policy arrangements to consolidate agricultural communes, but Mao Zedong castigated it for being made up of "women with bound feet" and suggested that "the rural socialist storm" was coming soon.

The development of agricultural production communes was advancing, a large number of elementary communes became advanced ones, and many mutual aid groups and individual farmers were directly brought into advanced communes. By the end of 1956, the number of agricultural production communes had grown to 750,000 (210,000 and 540,000 elementary and advanced communes, respectively), and the advanced communes accounted for 88% of all farming households nationwide. Approximately 600 million farmers had joined the communes, and the socialist transformation of rural areas in the production of materials was basically completed.

After his triumph with the collectivization movement, Mao Zedong believed that the transformation of the ownership system would bring about major development in terms of agricultural productivity, and in 1958, he proposed the general route, "Go all out, aim high, and achieve greater, faster, better, and more economical results" (*gu zu gan jin, li zheng shang you, duo kuai hao sheng* 鼓足干劲，力争上游，多快好省). He also argued that agricultural collectivization could only advance and that it was impossible to "retreat." He went even further and launched the Great Leap Forward and the People's Commune Movement, thus leading China's hundreds of millions of farmers to conduct an unprecedented and one-of-a-kind experiment. For workers, farmers, soldiers, educators, and merchants, politics and the commune were to be merged as one (*zheng she he yi* 政社合一), and the communes were to be "large in size and collective in nature" (*yi da er gong* 一大二公). People were supposed to eat without asking for money. The boundaries of communes, villages, and counties were to be overcome through teamwork. Organizations were militarized, actions became combat-like, and lives were collectivized (*zuzhi junshihua, xingdong zhandouhua, shenghuo jitihua* 组织军事化、行动战斗化、生活集体化). Social welfarization policies, such as the promotion of public canteens, were implemented. These utopian actions led directly to the unnatural deaths of 37 million people from 1959 to 1961.

In 1962, confronted with the abnormal conditions in the countryside, Mao Zedong had to personally establish the "Sixty Articles on People's Communes," which clearly specified that the production team was the basic ownership unit of the means of production (including rural land), as well as the basic accounting unit, and this was used to stabilize agricultural production relations. Since that time, though the rural system underwent many adjustments during the collectivization period, the basic framework of the system, as described in the slogan "Three-level Ownership, with the [Production] Team as Basis," has been maintained; the production team became the basic unit of accounting in production and distribution.[4]

The agricultural production communes were an important form of economic organization by which Mao Zedong guided peasants as they embarked on the socialist road. Based on the degree to which the means of production were held in common, these communes could be divided into elementary agricultural production communes, which were semi-socialist in nature, and advanced agricultural production communes, which were fully socialist in nature.

Elementary agricultural production communes were also called land cooperatives. They were still established primarily on the basis of private ownership of the means of production; members used land to become shareholders, and they had unified operations. Farm animals, large and medium-sized agricultural machinery, and other means of production were used in a unified manner in the commune, and members participated in the labor within the commune. After deducting annual production expenditures, taxes, provident funds, and public welfare funds, the surplus was divided among commune members as compensation for their labor, land, and other means of production. Compensation for labor took the form of labor division. Besides participating in labor within the commune, members could also cultivate their private reserved plots and operate family sideline businesses. The production tools of household sideline businesses, scattered trees, livestock, poultry, and materials for livelihood all belonged to the commune members. Although the land and other major means of production in these elementary communes were still privately owned, they were already practicing unified operation and were accumulating public assets.

Advanced agricultural production communes, or advanced communes, had achieved public ownership of land and other major means of production, and the distribution of consumer goods for members was based on labor. Members' privately owned land was converted without payment into collective ownership; their privately owned livestock and large- and medium-sized agricultural machines and implements, however, were purchased at cost by the commune or converted into collective property. Members' materials for livelihood, scattered trees, livestock, poultry, small farm implements, and tools needed for family sideline businesses were still privately owned by members. Advanced commune members took part in the labor within the commune. After deductions for taxes, production expenses, provident funds, and public welfare funds, the surplus left over from the advanced commune's total income was distributed among the members.

In the elementary communes, because the villages still owned the means of production and were protected by collective dividends, they usually worked in the farmland and used the oxen, boats, windmills, and other agricultural implements more responsibly. Since the villagers usually worked with other members of the elementary commune and supervised one another, they would pay attention to the quality of their farm work and use tools properly, even when they were working on land that did not belong to themselves or were using other people's tools. As a group, the villagers understood which land, which boat, or which cattle belonged to whom, so they would have to be a bit mindful when using them to ensure that the property was properly maintained and cared for. Even if someone used private property irresponsibly and was sloppy about his work, they could still use various collectively acknowledged form of punishment, such as public scolding, fines, or even expulsion from the commune. In general, members of commune living in the same village were usually connected by blood relations or friendships, so their cohesiveness was rather strong. These informal rules of the system guaranteed the performance of the elementary commune.

After the transition to the advanced communes, villagers lost their land and tools. They perceived cadres as actual employers who were controlling all of the collectives' resources. Members worked only for work points, paying attention only to the quantity of work done, not the quality. They no longer regarded the farmland and tools as their own property and no longer paid close attention to one another's work performance and tool usage as they had before. The quality and the efficiency of the farmwork thus deteriorated. Unlike the earlier elementary communes, which were established on the basis of village or neighborhood relations, the advanced communes generally included several natural villages separated from one another by a few kilometers. Whereas members of the elementary communes were connected to one another through kinship or as neighbors and could rely on a broad consensus on identity and interests that bound each to the other, the members of advanced communes were likely not to know one another because many did not live in the same village. It was precisely because of the lack of intimate ties among members of different villages that it was difficult for them to develop a sense of identity in the new collective organization. Since the advanced communes usually comprised several natural villages each, and the leadership of each comprised only a few individuals, it was impossible for them to distribute tasks to hundreds of members or personally supervise their labor. Therefore, given the failure of informal rule constraints and the difficulty of implementing formal rules, the poor performance of the advanced communes was a matter of course.

During the People's Commune era, the system of "Three-level Ownership, with the [Production] Team as Basis" (*san ji suo you, dui wei ji chu* 三级所有，队为基础) was implemented, and the production team became the basic accounting unit. In most cases by far, the production teams were equivalent to the original elementary communes. Each production team decided what kind of labor its members would do, how to supervise them, and how to record the work done. However, the production teams' property rights were not independent and exclusive, and its production arrangements and results were subject to the "three levels of ownership" (*san ji suo you* 三级所有). Whereas the elementary communes had been financially independent, the advanced communes practiced egalitarianism of distribution, which led to a weakening of the correspondence between how much effort that members put in and how many rewards they received. The correlation between farmers' labor input and income received from the collective declined.

In the commune and subsequent production team system, there were two other issues that plagued the communes' development and survival, besides operational and organizational efficiency, egalitarianism in distribution, and other systemic factors. One was that change carried out in the system through a top-down push would result in involuntary coercion and unwillingness among the farmers. The farmers were in a zero-sum game and had no right to withdraw, so they could only reduce their effort and respond by not cooperating, which then reduced the commune's performance. The other issue was the commission/agent problem of village cadres. After the transition to the advanced commune, the farmers lost all their land and large tools. They could no longer withdraw from the commune, and they were fully dependent on the cooperative for survival. Meanwhile, the commune cadres' attitudes toward members changed: they gradually turned toward "mandatory orders" whenever they carried out their superiors' orders or directives. Since they had various privileges, the commune cadres also distanced themselves from the masses, so any dereliction on their part in handling commune affairs and managing accounts became even more unresolvable. As a result, problems such as cadres amassing a surfeit of privileges, enforcing "mandatory orders" on members, mismanaging commune finances, and so on became ubiquitous. The communes fell into difficulty and eventually failed.

1.2 Establishment of the Household Contracting System

The Third Plenary Session of the Eleventh CPC Central Committee of the Communist Party (1978) opened the curtain on China's reform and opening up, of which the reform of the rural land contracting system led the way. This systemic change, which has had a major impact on the fate of hundreds of millions of farmers in China and the country as a whole, was initially launched in poverty-stricken areas. It later won the praise of reform-minded local and central leaders and was then carried out throughout the country through a policy push and finally through institutional establishment by law.[5]

In September 1980, when the CPC Central Committee issued a circular entitled "Several Problems in Further Strengthening and Improving the Responsibility System in Agricultural Production" (《关于进一步加强和完善农业生产责任制的几个问题》), it could only stress that "the collective economy is the unshakeable foundation of the advances of China's agriculture toward modernization," due to the constraints of the institutional environment and ideology of that period. At the same time, it required that China "should begin from [consideration of] actual needs and conditions and allow multiple forms of operation, multiple kinds of labor organizations, and multiple kinds of compensation methods to exist simultaneously." As for the various forms of the household contract responsibility system, such as the contracting of production to households (*bao chan dao hu* 包产到户, that is, assigning production responsibility to peasant households) and the contracting of wide-ranging output quotas to peasant households (*bao gan dao hu* 包干到户, which came later), these could only be subject to differential handling: they were limited mostly to remote mountainous regions and impoverished areas that had, for a long period of time, "depended on food sold back to them by the state in order to eat, depended on loans in order to produce, and depended on relief in order to survive," and to production teams where the masses had lost confidence in the collective.

By 1982, policymakers had become even more determined to implement reforms. The first central document of the CPC Central Committee for that year was devoted to the issue of agricultural production responsibility. This document emphasized that "China's agriculture must adhere to the road of socialist collectivization, and the public ownership of basic production materials such as land is, for the long term, unchanged." At the same time, it stated clearly that "the production responsibility system that the collective economy must establish is also, for the long term, unchanged." It clearly affirmed that the operational method of contracting of wide-ranging output quotas to peasant households, wherein each household operated independently and was responsible for its own profits and losses, was established on the foundation of public landownership. Farmers and collectives maintained a contractual relationship, which differed from the small, private, individual economy before communization, and they became an integral part of the socialist agricultural economy. In theory, it demonstrated the relationship between the agricultural production responsibility system and mainstream ideology and eliminated the ideological hindrance to the reform of the household responsibility system.

In 1983, the CPC Central Committee issued the circular, "Several Issues Concerning Rural Economic Policies (《当前农村经济政策的若干问题》)," which comprehensively demonstrated the rationality of the contract responsibility system for joint production:

Using farm households or groups as contracting units, it has expanded the autonomy of farmers, given full rein to the strengths of small-scale operation, and overcome the ills of overconcentrated management, ineffective shows of labor, and egalitarianism; it has also inherited the positive results of earlier communization and firmly maintained public ownership of basic production materials such as land as well as certain functions of unified management, which have enabled the new productivity to work more effectively for many years. This operational method, combining decentralized and unified operational methods, has broad adaptability; it can adapt to the current situation of manual labor and characteristics of agricultural production, and it can adapt to the need for productivity development in the process of agricultural modernization. Under this mode of operation, household management contracted by separate households is only one management level in the cooperative economy and a new type of family economy. It is essentially different from the small, private, individual economy of the past, and [the two] should not be confused.

From this time, the debate at the policymaking level on the basic system for rural areas came to an end, and the household contract responsibility system was universalized throughout the country. By the end of 1984, 99% of the production teams and 96.6% of the farmers in China were engaged in the contracting of output quotas to farm households.[6]

1.3 Formation of the Household Contract Responsibility System

With the establishment of the rural household contract responsibility system, the policy orientation of the rural system mainly focused effort on empowerment with property rights, perpetuation, legalization, and institutionalization of the right to manage agricultural land contracting.

First was to continuously extend the land contract period and stabilize the relationship between farmers and the land. The first central document of CPC Central Committee was issued in 1984, "Circular on Rural Work in 1984" (《关于1984年农村工作的通知》; CPC Central Committee and State Council, 1984, No. 1) proposed "extending the land contract period, encouraging farmers to increase investment, cultivating land resources, and implementing intensive operation. The land contract period should generally be more than 15 years." During the second round of contracting in the early 1990s, the land contract period was extended to 30 years. In 2008, the Third Plenary Session of the Seventeenth Central Committee proposed "stabilizing the existing land and contracting relationship and implementing it over a long term without change."

Second was to establish the household contract responsibility system as the basic economic system of rural China. In 1991, the CPC Central Committee, in its "Decisions on Further Strengthening Agriculture and Rural Work" (《关于进一步加强农业和农村工作的决定》), first proposed: a two-tier management system that combines the responsibility system based primarily on household contract production and a system of unified management combined with independent management, to serve as the basic system of rural collective economic organizations in China, to be stabilized over the long-term stability and continuously improved.

This was given further emphasis in "Policy Measures Concerning the Current Development of Agriculture and the Rural Economy" (《关于当前农业和农村经济发展的若干政策措施》; CPC Central Committee and State Council, 1993, no. 11) and the Third Plenary Session of the 15th CPC Central Committee in 1998. The 1999 *Constitution* clearly states, "The rural collective economic organization implements a two-tiered management system based on family contract management and a combined unified and independent management." "The rural collective economic organizations apply the two-tiered management system characterized by the combination of centralized operation with decentralized operation on the basis of operation by households under a contract." *The Law of the People's Republic of China on the Contracting of Rural Land* of 2002 clearly states that "the state implements the rural land contract management system." In 2003, the Third Plenary Session of the 16th CPC Central Committee further strengthened the expression: "Land household contracting management is the core of the basic rural management system." *The Real Right Law of the People's Republic of China* of 2007 stipulated:

The rural collective economic organization shall implement a two-tier management system characterized by the combination of centralized operation with decentralized operation on the basis of household contracted management. The system of land contracted management shall be implemented to the cultivated land, wood land, grassland, and land for other agricultural uses that are owned by farmers' collectives as well as those that are owned by the state and exploited by farmers' collectives.

When this system came under continuous questioning, the Third Plenary Session of the 17th Central Committee in 2008 fully affirmed the status and function of this system with a very clear attitude:

The two-tier management system based on household contract management and a system of unified management combined with independent management is the basic rural management system, one that adapts to the socialist market economy system and conforms to the characteristics of agricultural production. It is the cornerstone of the party's rural policy and must be unswervingly upheld.

Third was to clarify the content of land rights under the household contract system and to improve land contract management rights.

(a) *The public nature of farmland was declared.* At the beginning of reform, it was made clear that "contracted land may not be bought or sold, may not be leased out, and may not be converted into a homestead or land for other non-agricultural uses." When *the Law of the People's Republic of China on the Contracting of Rural Land* was promulgated in 2003, it was further clarified legally: "After the contracting of rural land, the nature of the land's ownership remains unchanged. The contracted land cannot be bought or sold."

(b) *The relationship between the collective owner and the contractor was clarified.* The *Law of the People's Republic of China on Land Administration (Adopted at the 16th Meeting of the Standing Committee of the Sixth National People's Congress on June 25, 1986)* states, "Land owned by farmers' collectives shall be contracted and managed by members of the collective economic organization. The contractor and subcontractor shall conclude a contract that stipulates the rights and obligations of both parties." *The Law of the People's Republic of China on the Contracting of Rural Land (Adopted at the 29th Meeting of the Standing Committee of the Ninth National People's Congress on August 29, 2002)* stipulates, "Members of rural collective economic organizations have the right to subcontract rural land contracted out by the collective economic organization according to law. No organization or individual may deprive or illegally restrict the right of members of rural collective economic organizations to subcontract land."

(c) *The farmers' contract management rights were protected as property rights.* Policies and relevant laws through the years stipulate that during the period of statutory contracting, no organization or individual may interfere with farmers' autonomy in production and operation, illegally adjust and take back contracted land, forcibly transfer contracted land against the farmers' will, or illegally occupy farmers' contracted land.

(d) *The power of land contracting right was improved, and land property rights of farmers was protected by law.* The Third Plenary Session of the 15th CPC Central Committee (1998) and the Third Plenary Session of the 17th CPC Central Committee (2008) proposed to protect farmers' rights to occupy, use, and derive income from contracted land in accordance with law, and to grant farmers long-term and guaranteed land use rights. During the statutory contracting period, no organization or individual may interfere with the farmers' autonomy in production and operation, illegally adjust and take back contracted land, forcibly transfer contracted land against the farmers' will, or illegally occupy farmers' contracted land. *The Law of the People's Republic of China on the Contracting of Rural Land (Adopted at the 29th Meeting of the Standing Committee of the Ninth National People's Congress on August 29, 2002)* further listed out the rights of contractors, that is, the right to use, derive income from the contracted land, the right to transfer the contracted land management right, and the right to autonomously organize production, operation, and disposal of products. If the contracted law is expropriated or occupied, farmers have the right according to law to obtain corresponding compensation. During the contract period, the contractor may not take back the land from the subcontractor nor make adjustment to the subcontracted land.

(e) *Special provisions were made to protect of subcontracting rights and interests of migrant workers.* In response to cases in which farmers in many places lost their subcontracting rights when they went elsewhere for work, and an increase of disputes arising from such cases, the State Council in 2004 issued the "Urgent Notice on Appropriate Resolution of Current Rural Land Contract Disputes (《关于妥善解决当前农村土地承包纠纷的紧急通知》)," which clearly stipulated:

> During the statutory contracting period, no organization or individual may interfere with the farmers' autonomy in production and operation, illegally adjust and take back contracted land, forcibly transfer contracted land against the farmers' will, or illegally occupy farmers' contracted land. It is necessary to respect and protect the land contracting rights and operational autonomy of migrant workers. During the contracting period, except in cases when the entire family of the contracting party change to a nonagricultural household register upon

moving into a city divided into districts, the farmer's land contractual management right shall not be rescinded. Farmers who return home to work in agriculture, provided that they have obtained the contracting right during the second round of contracting, must return the contracted land to the original contracted farmers to continue farming. If the rural organization has already contracted out, on a short-term contract, to another farmer the land of the migrant worker, then it must give the income derived from that contract to the original farmer who owns the land contracting right, and after that [short-term] contract expires, he must return the land to the original contracted farmer for cultivation. If it is a long-term contract, the contract may be revised and the contracted land must be returned to the original contracted farmer in a timely manner, or, on a consensus basis, resolved through payment or increase of compensation to the original contracted farmer. As for the small number of farmers who have gone out for work elsewhere and did not participate in the second round of contracting, but who have now returned to the countryside demanding contracted land, it is necessary to distinguish their different situations and properly handle the cases through democratic consultation. If the farmer's household registration is still in the rural village, in principle it should be agreed that he should continue participating in land contracting. If conditions permit, [the case] should be settled by an adjustment of mobile land;[7] if there is no mobile land, [the case] can be settled through land transfer.

When the Law of the People's Republic of China on the Contracting of Rural Land was enacted in 2002, legal provisions were also made for the disposal of contracted land after farmers went out to work elsewhere:

Within the duration of the contract, if the household that undertakes the contract shifts to a small town and settles down there, its right to operate the contracted land shall be sustained, or its lawful right to circulate the operation right of the contracted land shall be permitted according to its free will. Within the duration of the contract, if the household that undertakes the contract shifts to a city that is divided into regions, thereby becoming a nonagricultural household, the cultivated land and the contracted grassland shall be returned to the party that gives out the contract. If the household undertaking the contract refuses to return such land, the party that gives out the land shall be entitled to withdraw it. Within the duration of the contract, when the party undertaking the contract returns the contracted land to the party that gives out the land or when the party that gives out the contract lawfully withdraws the contracted land, if the input of the party undertaking the contract improves the production capability of the contracted land, it shall deserve a corresponding compensation.

(f) *A registration system was established for land contractual management rights.* The *Measures of the People's Republic of China for the Administration of the Certificates of the Right to Contracted Management of Rural Land* (《农村土地承包经营权证管理办法》), enacted in 2004, requires the local people's governments at or above the county level to issue certificates for the right to contracted management of rural land. *The Real Right Law of the People's Republic of China* (2007) stipulates:

The right to the contracted management of land shall be established as of the effectiveness of the contract on the right to the contracted management of land. The local people's government at or above the county level shall issue a certificate of the right to the contracted management of land . . . and register it in the brochure and confirm the right to the contracted management of land.

The right to the contracted management of land shall be established as of the effectiveness of the contract on the right to the contracted management of land. The local people's government at or above the county level shall issue a certificate of the right to the contracted management of land, a forestry right certificate or a grassland-use right certificate to the holder of right to the contracted management of land, register it in the brochure, and confirm the right to the contracted management of land.

The first central document of the CPC Central Committee (2008) demanded the necessity to "ensure rural land contracted management rights to the households." The first central document of the CPC Central Committee (2008) (2009) demanded: ". . . steadily implement the pilot project for the registration of contracted management rights of land and carry out the certification of the area, space location and ownership of the contracted land parcels to the farmers." The first central document of the CPC Central Committee (2008) (2010) reiterated, ". . . fully implement the 'four to the household' [measures] of contracted land parcel, area, contract, and certificate, and expand the scope of the pilot program to register the contracted management rights of rural land." The first central document of the CPC Central Committee (2008)

(2012) demanded, "Accelerate the rural cadaster surveys and have the confirmation, registration, and certification of ownership of the various categories of rural collective land basically completed in 2012."

(g) *The system for transferring contracted management rights for farmland was improved.* The first central document of the CPC Central Committee (2008) (1984) already stipulated:

When a collective member requests contract termination or reduction of contracted land because he is no longer able to cultivate it or has transferred to a different industry, the land may be handed over to the unified arrangement of the collective, or with the consent of collective members, negotiate the transfer of the contract to another party, but they cannot modify the content of the terms of the contract with the collective. The subcontracting conditions may be agreed upon by both parties, based on local conditions. The farmers' investment in land should be reasonably compensated. Some specific methods, for example, on giving grades to or fixed evaluations on land, may be formulated through members' democratic consultation, which may then be used as a reference for compensating farmers on investment at the time the land use rights are transferred. Reasonable compensation measures should also be stipulated for reductions of land productivity due to predatory management.

Documents from the Third Plenary Session of the 14th CPC Central Committee in 1993 stipulated:

With the insistence on the collective ownership of land as a premise, the contracted management right of inherited development production projects is allowed, as well as the compensated transfer of the land use rights in accordance with the law. In a small number of places where the economy is relatively developed, various forms of moderately scaled management for transfer of contract, shareholding, and so on may be adopted, based on the principle of voluntary participation of the masses.

In 1998, the Third Plenary Session of the 15th CPC Central Committee stipulated:

The rational transfer of land use rights must be carried out in accordance with the principles of voluntariness and compensation, and the transfer cannot be forced on a farmer for any reason. In a small number of places that indeed meet conditions, various forms of moderately scaled operations of the land may be developed, on the basis of increasing the degree of agricultural intensification and of the voluntary participation of the masses.

By 2008, the Third Session of the 17th CPC Central Committee had already fully updated the content:

Strengthen the management of and servicing of rights to the contracted management of land, establish and improve the market for the transfer of land contractual management rights, and allow farmers to develop various forms of moderately scaled operations for the transfer of land contractual management rights, such as transferring contracts, leasing out, exchanging, shareholding, etc. Where conditions permit, it is possible to develop business entities of scale such as large professional households, family farms, and farmers' professional cooperatives. The transfer of land contractual management rights shall not change the collective ownership of the land and the use of the land or damage the land contractual rights of the farmers.

The first central document of the CPC Central Committee (2008) (2009) states, "Adhere to transfer the land they contract to work on in return for compensation on a voluntary basis and in accordance with relevant laws, and respect the farmer's principal status in land transfer. No organization or individual may force a transfer or prevent a voluntary transfer."

The first central document of the CPC Central Committee (2013) states:

> Adhere to transfer the land they contract to work on in return for compensation on a voluntary basis and in accordance with relevant laws, guide the orderly transfer of rights to the contracted management of rural land, encourage and support the transfer of contracted land to large professional households, family farms, and farmers' cooperatives, and develop various forms of moderately scaled operations. In combination with the basic construction of farmland, farmers are encouraged to adopt mutually beneficial exchange methods to solve the problem of the fragmentation of contracted land. Land transfer must not be done through mandatory orders, to ensure that farmers' rights and interests are not harmed, that the land use is not changed, and that the land's general agricultural productive capacity is not destroyed. The establishment of a strict system for permitting access to and monitoring farmer-contracted farmland (forest land, grassland) leased out to industrial and commercial enterprises is to be explored. Standardize land transfer procedures, gradually improve third-level service networks in counties and villages, and strengthen services for communicating information, policy consultation, contract signing, price evaluation, and so on.

1.4 Features and Problems of the Current System for Land Contractual Management Rights

Taking stock of the salient features of current system for land contractual management rights in rural China, we may identify the following:

(a) *Firm adherence to the collective ownership of land.* The collective ownership of land is the most important legacy of the movement in China to communalize rural land. It is a major component of China's basic, socialist system and the single institutional arrangement that most distinguishes China's rural system from those found in other countries and regions in the world. Adherence to the collective ownership of land was the premise and foundation for the CPC's attainment of power, and it also became the basis that influenced the changes in China's rural system. Therefore, even though it had encountered three years of unremitting disasters during the collectivization period, the bottom line of the rural system returned only to the kind of collective ownership characterized by "Three-level Ownership, with the [Production] Team as Basis" (*san ji suo you, dui wei ji chu* 三级所有，队为基础). During the surging wave of reforms in the 1980s, when reformers were promoting management by households and reinstating the system of landownership by farmers, they also repeatedly emphasized that the household responsibility system was being carried out under the premise of collective ownership. During the 1990s, when the basic management system for rural areas was being defined, it was again necessary to emphasize "unified management combined with independent management" (*tong fen jie he* 统分结合), with collective ownership being the most important source of legitimacy in the "unified" part of the agricultural management system. Each collective member equally enjoys the right to use collective land, and collective ownership evolves into a collective ownership of membership. Each member of the collective enjoys the right to distribute income after the land is no longer farmed.

(b) *Farmers become owners of land property rights.* The most important institutional aspect of rural reform is that it gives land property rights to the contracting farmers, thereby solving the problem of low incentives for agricultural production and of institutional instability through the protection and enforcement of property rights. Under the collective ownership system, farming households have replaced the production team as the decision-making entities of agricultural production and economic activities, and they possess the contracting management right, income right, and subcontracting right for their farmland during the contract period. In recent years, farmers have acquired land property rights as usufruct through the improvement of land property rights, which included the clear identification and consolidation of the prime position of the farming household, the stabilization of land contracting relationship (extending duration of contract from 15 to 30 years to perpetuity), and clearly putting farmers the main actors in arranging for land transfers. Meanwhile, in the granting of property rights, the land rights obtained by subcontracting farmers are a combined right to the contracted management of land, that is, the right to subcontract and the right of management.

(c) *Family management is the main form of business and organization in agricultural production.* Replacing the production team form of management from the collectivization period with family management was one of the important achievements of the rural system reforms. Family management has become the most important organizational and business form in agricultural production, and it was determined by the characteristics of agricultural production and labor. Agriculture demands of its producers that they take very close care of all its aspects, while labor demands a direct correlation between the effort put in and the returns. Therefore, as long as the natural risks of agriculture are not eliminated, agricultural investment still depends on labor investment, and family management is the most appropriate organizational form of agricultural management. At present, in the debate over the basic management system in rural areas, family management is the most heavily criticized—the voices denigrating family management are getting louder and louder, so we need to get more serious in responding to them. In our view, what the reform of agricultural management form needs to accomplish is not to denigrate the family-based organization but to give it new meaning and power.

However, through its operation, the current system of rights to the contracted management of rural land also shows its systemic shortfalls. The inability of this system to adjust, especially to accelerating urbanization and the increasing number of people who no longer work in agriculture, has gradually become apparent.

(a) *The collective ownership of membership rights continues to strengthen itself.* In 1998, a survey of 800 households showed that, with regard to the central government's emphasis and implementation of the "policy on the extended 30-year contracting period," 62% of those interviewed did not approve of the extension of the contracting period to 30 years, during which no land adjustment could be made, and in the more traditional the region, the higher the percentage. Moreover, 81% of the respondents did not approve of "the method of not allocating land to newly added members."[8] By 2003, when the *Rural Land Contract Law* was implemented, a survey of more than 2,000 households showed that even though the proportion of farmers who approved of the extension of the contracting period to 30 years had risen to 62.9%, another 20.6% of the farmers clearly advocated shortening the period. Those approving the non-adjustment of land during the contracting period had risen to 51.1%, but 36.8% still thought that the land could be adjusted during that period.[9] By 2008, a survey of 2,200 households showed that the proportion of respondents who thought that not adjusting the land at all during the 30-year contracting period was unreasonable was as high as 62.79%. The proportion of respondents who believed that "not increasing the land for an increase in the number of people" or "not reducing the land for a decreased number of people" were as high as 61.98% and 59.95%, respectively.[10]

(b) *There are problems in rights protections after the right to contract land gets separated from the right of management.* China's current laws clearly protect farmers' land contract management rights. However, these rights are combined in terms of their setup. Within the context of high-speed industrialization and urbanization, the mobility of the rural population and labor force has also accelerated, so that the trend wherein rural land contracting rights and management rights are separated is becoming increasingly obvious. Nationwide, the number of migrant workers who currently spend six months or more working away from home has reached 250 million. These people continue to have land contracting rights, but the land management rights have been controlled by the elderly and women who remain at home, and in some cases, they entrust the management of the land to neighbors and relatives at a very low price, or even for free. Field research in some locales has found that this separation between contracting rights and management rights is becoming even more obvious. In Shanghai's Songjiang District, agricultural labor in 2011 accounted for only 2.9% of the labor force, but nearly two-thirds of its registered rural population was residing in cities, which indicates how, in suburban areas near cities, the land management rights have basically been separated away. In highly industrialized areas such as Guangdong, Zhejiang, and Jiangsu, the proportion of local rural laborers who are actually engaged in agriculture is less than 10%. This represents how, in developed areas, the land contracting rights have to a large extent been separated from the management rights. This trend is very obvious even in rural areas in Central China. From 2000 to 2010, the number of rural residents in Anhui decreased by 9.3 million people, while the proportion of rural-registered persons in cities increased. In Feidong County (肥东县), during one nearly two-year period, rural buyers accounted for 67% of housing sales.

The huge changes in the distribution of the rural population and the labor force have brought about a de facto separation between farmers' land contracting rights and their land management rights. Long-term migrant workers and farmers who have purchased houses in urban areas continue to have the right to contract land, but they no longer manage the land. This phenomenon of farmers who have land contracting rights but who do not necessarily continue to manage the land is becoming more and more common. As a consequence, the concept and content of "contractual management rights" as generally proposed in law faces difficulties when implemented in the real world. The problem is specifically manifested this way: on the one hand, protecting the contracting rights can lead to a weakening of the management rights, which is not conducive to the circulation of rural land and the moderate scale expansion of land management; on the other hand, as soon as management rights are emphasized, the policy orientation will tend to increase the circulation of rural land, which in turn will lead easily to the loss of contracting rights by members of the original rural collective organizations. At present, and in the near future, it will be a huge challenge to determine how effectively to protect the land contracting rights of the members of the original rural collective organizations, and at the same time legally protect their land management rights, in order to achieve the goals of guaranteeing farmers' land rights in accordance with the law, promote the circulation of rural land, and modernize agriculture.

(c) *In the set-up of the right to contracted land, certain powers are deficient.* Judging from the establishment and empowerment of the household contracting system, the policies and laws so far have focused mainly on protecting the farmers' land contractual management rights, that is, on strengthening and protecting farmers' rights to use and derive income from farmland, but there has not been enough emphasis on granting farmers rights in other respects.

The first is the right of disposal. No matter how much the property nature of the contracted land is emphasized, under the premise of collective ownership, the contracting farmers still cannot buy or sell the contracted land. As the population urbanizes and increasingly pursues nonagricultural work, some of the urbanized farmers face the problem of how to dispose of their contracted land, and as more and more farmers settle in the cities, this issue becomes more and more prominent. The current law stipulates that when farmers settle in large- or medium-sized cities, they must return the contracted land to the collective. This provision obviously deviates from the principle of protecting land as property. Therefore, if the right to dispose of the contracted land is not in place under the collective ownership of land, this can lead to a violation of farmers' land property rights.

The second is mortgage and security rights. As a property right, the land contracting right should be whole bundle of rights. However, within the set-up of the system that we have described above, wherein the combined rights of contracting and management are protected, the mortgage and security rights that are granted to holders of land contractual management rights are, in practice, fraught with greater risk. If the contractual management right is mortgaged or used as security, and the contractor uses his contracted land as security for obtaining financial support, the powers and responsibilities are equivalent, and the bank will not have any problems in operation. However, if the contractor transfers the management right to a new manager, the latter has only the management right—not the contracting right. If the latter uses this management right to get a mortgage for a loan and then cannot repay it due to some management issue, then the bank has to take possession of the land—it cannot take possession of the management right. In this way, the original contractor faces the risk of losing his contracting right. Therefore, this design defect in the contractual management right itself has also led to a lag in empowering farmers with mortgage and security rights.

The third is the right of inheritance. Not long after the implementation of the policy of "an increase in the number of people [in a household] will not entail an expansion of the land, and a decrease in the number of people will not entail a shrinking of the land" (zeng ren bu zeng di, jian ren bu jian di 增人不增地，减人不减地), and the policy of "over a long term without change" with regard to contracting period, the land contracting system became a land system without any future end-date. The contracted land in the farmer's hands, as property, became fixed. After the "over a long term without change" policy was implemented, one of the problems that must be faced is how to deal with the contracted land as the number of persons in any given household inevitably changes in the future. At present, neither law nor policy touches on the problem of how to deal with such contracted land—but the problem must be faced. Specifically, once the person holding the contract passes away, the contractor's contract is passed to the next generation or returned to the collective. Even if it is passed on to the next generation, shall the land continue to be divided equally among the children or passed down only to the eldest son (as in Japan and some other countries), to prevent the land from getting divided into ever-smaller plots? At present, the age of contractors from the first generation of contractors in China is already close to average life expectancy, so the issue of inheritance rights for contracted land is already before us, and it requires an institutional response.

2. The Homestead System: Features and Problems

2.1 Formation of the Homestead System

The homestead system is a very unique institutional arrangement of the Chinese land system. In countries that follow the traditional continental law system, there are provisions for territorial rights but nothing about homesteads. In China, people through history have generally referred to their homesteads as the "house" (*zhai* 宅, *zhai zi* 宅子, or *fang zhai* 房宅). These synonymous terms included both the house that a family had built and of course the land that it occupied.[11] In more than 60 years that have elapsed since the founding of the new China, when China's rural countryside was being put through a multitude of reforms and changes—the cooperative movement, the People's Communes, the household contracting system—the homestead system was also being put through many changes, which were basically similar to and yet different in terms of their institutional arrangements from those of the rural land contractual management rights system. All this highlights the more complex nature of the homestead system and the difficulties of reform to that system.

1. *The era of land reform and agricultural cooperatives: private ownership of homesteads.* During and after the establishment of the new China, the land reform that took place was aimed at "changing feudal and semi-feudal land ownership step-by-step into peasant land ownership," and the protection of private property, including that of farmers, was explicitly proposed. After land reform was completed, the people's government specially issued landownership certificates to farmers. The first *Constitution* of the new China in 1954 clearly stated, "The state protects the right of farmers to own land and other means of production in accordance with law," and, "The state protects the rights of citizens to own lawfully-earned income, savings, houses, and other means of life." As part of the farmer's privately owned land, the homestead was naturally protected, equally and legally.

   Starting in 1953, China's rural areas entered a period of socialist transformation. During the elementary commune stage, farmers' land began to be managed in a unified manner, through shares, but it remained privately held. As for the homesteads, the *Model Regulations for Advanced Agricultural Producers' Cooperatives* of March 1956 stipulated that "[commune] members' original tomb grounds and house foundation land are not to be taken into the commune. The foundation land required for the construction of new houses and tomb grounds as required shall be decided as a whole by the commune, and when necessary, the commune may apply to the township People's Committee for assistance." The *Model Regulations* were the first law with provisions concerning homesteads. During this period, even though the land underwent a transformation from private ownership by farmers to collective ownership by communes, the farmers' houses remained retained their private attribute. The term "homestead" (*zhai ji di* 宅基地) did not appear directly in the laws of this period, nor did the law strictly distinguish between the land that the house occupied, the house built on it, and other attachments. Therefore, the farmer's ownership of the house simultaneously meant his ownership of the homestead—the land occupied by the house. Meanwhile, the right of farmers to buy, sell, mortgage, and lease houses was protected.

2. *The People's Commune era: homesteads are collectively owned but used by the farmers, and the houses are privately owned.* The *Draft Amendment of the Regulations Governing Rural People's Communes* of 1962 contained but one simple sentence on homesteads: "Members' homesteads are categorically not permitted to be leased out, bought, or sold." Provisions on the rights appertaining to farmers' houses, however, were very detailed and extremely exclusive, such as "Members' houses are always owned by the members." Also:

   A member has the right to buy, sell, or lease a house; a member leasing out or selling a house can negotiate a fair and reasonable rent or house price through an intermediary, and both parties in a sale or lease shall sign a contract. No unit or individual is permitted to force a member to move, and no rent or price is to be paid without the member's agreement. No agency, group, or unit may occupy a member's house. If the house of a member must be requisitioned due to construction or other needs, the State Council's regulations on the requisitioning of houses should be strictly followed, compensation given, and resettlement properly sited.

These *Regulations* not only used the term "homestead" for the first time in China's policy documents and laws, but also created the institutional arrangement separating the homestead from the house built on it.

The "Notice on Several Supplementary Provisions Concerning the Member Household Issue in Various Places," issued by the CPC Central Committee in 1963, emphasized, "A member's homestead is collectively owned by the production team and is categorically not permitted to be leased out, bought, or sold." At the same time, it clearly stated the right of use of the homesteads "belongs to the various members, long-term, and shall not be changed over the long term. The production team should protect members' use rights and cannot requisition or reallocate [homesteads] at will." As for the attachments to homesteads, "including the house, trees, sheds, pigpens, latrines, etc., we emphasize, shall permanently belong to the members, and the members have the right to buy, sell, or lease them." It further stipulated: "After the house is sold, the right to use the homestead is transferred to the new house owner, but ownership of the homestead still belongs to the production team." This "Notice" also proposed for the first time the method for obtaining a farmer's homestead:

When a member needs to build a new house and does not have the homestead, his household shall submit an application, and after the matter is discussed and agreed upon by the general assembly of members, the unified planning, assistance, and problem-solving shall be done by the production team. Whether or not the land occupied by the member's new house is cultivated land, no land price shall be paid.

We can see that this "Notice" confirms the separation between the house and the land it sits on, and for the first time it clearly confirms the separation of homestead ownership from the right of use of the homestead, the means of applying for a homestead, and the fact that it is to be obtained without payment.

Also in 1963, the Supreme People's Court issued the "Opinion on Several Issues Concerning the Thorough Implementation of Civil Policies (Revised Draft)" (《关于贯彻执行民事政策几个问题的意见》(修正稿)), which clarified provisions on the rights involved in farmers' housing transactions:

For any house that is permitted to be bought or sold according to the law, after a proper legal procedure has confirmed the purchase and sale relationship of the house, the rights of the two parties shall be protected, and neither party shall regret or rescind the sale and purchase contract. An individual citizen has the right to lease out a house belonging to that individual and collect fair and reasonable rent. The right of laborers to pawn their houses shall be recognized except in cases that have already been settled in land reform, and they shall be allowed to redeem them at the expiry of term.

The Supreme Court's finalized opinion was promulgated in 1979 when it issued the "Opinion on the Thorough Implementation of Civil Policy Laws" (《关于贯彻执行民事政策法律的意见》), which made specific arrangements for handling disputes over members' use rights on homesteads. This included the statement that the right to use the homestead is "based on the principles that landownership belongs to the state or the collective, and that the leasing out or sale and purchase [of the homestead] is categorically not permitted."

Also, wherever the ownership of the homestead as determined during the local land reform is changed to use rights, the right to use that homestead is unchanged; wherever the local homesteads have been uniformly planned, the member's use rights of the homestead shall be handled based on the determination of the plan; and wherever an adjustment has been made through a legal procedure, they shall be handled based on the adjustment decision. If a change to the rights to the homestead is necessary, the People's Court shall, in accordance with the policies and laws of the Party and the state, study and properly handle it with the relevant departments. Fruit trees and bamboo planted by members on the homestead are owned by the members.

By this time, the rudiments of the Chinese homestead system had basically taken shape. These were, first, collectively owned by the production team, and members were prohibited from leasing out, buying, or selling it. Second, ownership of the homestead was separated from the right to use it; the farmers possessed the rights of use for their homesteads over the long term, and these were protected by law. Third, the homestead was separate from any rights arrangements relating to the house on it. Fourth, farmers had exclusive ownership of their houses and could buy, sell, lease, mortgage, and pawn them. Fifth, homestead use rights are transferred with the purchase, sale, and lease of the house. Sixth, an application-based system for farmers to obtain homesteads free was established.[12]

3. The period of reform: construction of the homestead rights system. After reform and opening up, China's administration with regard to homesteads was put on the right track. A series of laws and guidance documents to strengthen the management of homesteads was issued, a sound administrative agency was established, and relatively clear regulations were formulated, thus creating a rather unique rights system in relation to homesteads.

First, ownership and rights appertaining to homesteads were restricted. In 1981, the State Council promulgated the "Emergency Notice on Stopping Rural Buildings from Encroaching on Cultivated Land" (《关于制止农村建房侵占耕地的紧急通知》; State Council, no. 38), which set out clearly: "As for the homesteads allotted to [collective] members, members only have the right of use, and they are not permitted to lease out, buy, sell, or transfer them without authorization, and they are not permitted to build houses, dig tombs, open mines, fire bricks and tiles, etc. on contracted land or private reserved plots." In 1982, the *Regulations on the Administration of Land Used for Housing Construction in Villages* (《村镇建房用地管理条例》) reiterated:

Members have no ownership of the homesteads, only the right to use them in accordance with the specified use. They may not build houses, dig tombs, open mines, destroy fields, fire bricks and tiles, etc. on private reserved plots, private reserved mountains, forage land, or contracted land. The purchase, sale, leasing out, and unauthorized transfer of construction land is strictly forbidden.

The 1982 *Constitution* and the 1986, 1988, and 1998 versions of the *Law of the People's Republic of China on Land Administration* all stipulate, "The homestead belongs to collective ownership," and, "To sell or lease out a homestead and then apply for a homestead shall not be approved."

Second, conditions were set on obtaining a homestead. The "Notice on Further Strengthening Land Management to Effectively Protect Cultivated Land" (《中共中央国务院关于进一步加强土地管理切实保护耕地的通知》; CPC Central Committee and State Council, No. 11) of 1997 proposed:

Residential housing for rural residents must strictly follow the standards set by the provinces, autonomous regions, and direct-administered municipalities, and the homesteads obtained according to law. Each rural household can have only one homestead not exceeding the standards. Excess on homesteads must revert to collective ownership according to law.

In 2004, the Ministry of Land and Resources issued the "Opinion on Strengthening Administration of Rural Homesteads" (《关于加强农村宅基地管理的意见》; no. 234), which stipulated:

[Set] strict conditions on homestead applications. Resolutely implement the "one house per household" legal provisions. One rural household can only own one homestead, and its area must not exceed the standards set by the province (autonomous region, municipality). All localities should, in light of local conditions, formulate uniform standards for the area of homesteads and conditions on applying for homesteads. An application shall not be approved if the conditions on applying for a homestead are not met. If the rural residents sell, lease out, or give their original housing to other persons and then apply for a homestead, the [application] shall not be approved.

Third, the relationship between the house and the land as well as the use rights for the homestead were determined. In 1989, the State Bureau of Land Administration issued "Several Opinions on the Issue of Determining Land Rights" (《关于确定土地权属问题的若干意见》; no. 73), which stipulated that property owners enjoyed the same land use rights of other members of the collective to which the land belongs. As for the homesteads of registered nonagricultural rural residents, where their ownership of the property had not changed, their right to use collective land for construction could be determined in accordance with law. For homesteads obtained through inheritance of a house, heirs had the use right. If an heir already had a homestead and the total area exceeded the prescribed standard, then the right to use the collective land for construction purposes could be determined temporarily according to the actual use area. If the total area of a homestead obtained through transfer or purchase of a house plus the [purchaser's] original homestead exceeded the standard set by the local government, then the right to use the construction land of the collective's land was determined according to the relevant regulations. For rural nonagricultural construction land and construction land apart from homesteads of rural registered professional households, the right to use collective land for construction was determined separately. As for the right to use construction land as rural homesteads, whenever a homestead exceeded the area standards set by the local government, then, when the house was later razed, altered, or rebuilt, the use right was redetermined based on the area standard fixed by the local government, and any excess had to be returned to the collective.

Fourth, the rights structure of homesteads was established. After continuous improvement, by the time that the *Real Right Law of the People's Republic of China* was promulgated in 2007, the system of rights relating to homesteads was formed. Specifically, the use rights holder of a homestead enjoys the rights of occupying and using the collective-owned land in accordance with law, and has the right to use that land to build a house and ancillary facilities in accordance with law. If the homestead is lost due to natural disaster, the right to use the homestead is eliminated. Rural residents who have lost their homesteads should be reallocated another homestead. If the registered use rights of the homestead have been transferred or destroyed, then a change of registration or a cancellation of registration should be processed in a timely fashion. The use rights of a homestead cannot be mortgaged. The "Decision on Several Major Issues on Promoting the Reform and Development of Rural Areas" (《关于推进农村改革发展若干重大问题的决定》), produced by the Third Plenary Session of the 17th CPC Central Committee, recommended, "Improve the rural homestead system, make the management of homesteads strict, and protect rural households' usufructuary rights on homesteads in accordance with the law." The first central document of the CPC Central Committee (2013) proposed, "Reform and improve the rural homestead system, strengthen its management, and protect rural households' use rights on homesteads in accordance with the law."

Fifth, the disposal of nonmembers' homesteads was worked out. In the 1982 *Regulations on the Administration of Land Used for Housing Construction in Villages* (《村镇建房用地管理条例》), it was stipulated:

Rural (collective) members, workers, and veterans who have returned to their hometowns after resigning or retiring, and overseas Chinese who have returned to settle in their hometowns and need a homestead for building a house should submit application to the production team, and after the members' general assembly discusses and approves it and the production team approves it, it is reported to the commune management committee for approval. If it is necessary to occupy cultivated land or orchards, it must be reported to the county-level people's government for approval. After approval, the approval authority shall issue the certificate of use of the homestead.

Both the 1986 and 1988 versions of the *Law of the People's Republic of China on Land Administration* stipulate that whenever registered urban, nonagricultural residents need to use collective-owned land to build housing, they need to obtain the approval of the county-level people's government, and the area of the land used may not exceed the standards set by the relevant province, autonomous region, and direct-administered municipality. They must also pay the land compensation fee and resettlement fee based on the state standards of construction land expropriation.

The *Regulation on the Implementation of the People's Republic of China on Land Administration Law* (《土地管理法实施条例》) of 1991 stipulated:

In the case when a registered urban, nonagricultural resident needs to use collective-owned land to build housing, then after obtaining the consent of his unit or residents' committee, he must submit an application for land use to the rural collective economic organization or to the village committee or to the village (township) collective economic organization where that land is located. Where the land to be used is owned by a rural collective, then, after the application is discussed and passed by the rural representative meeting or general assembly and has been reviewed and approved by the village (township) rural collective people's government, it is reported to the county-level people's government for approval. Where the land to be used is owned by a village (township) collective, then, after the application is discussed and passed by the village (township) economic organization and has been discussed and approved by the village (township) people's government, it is reported to the county-level people's government for approval. For staff and workers returning to their villages of original registry, and retired veterans, resigned or retired cadres, and overseas Chinese and compatriots of Hong Kong, Macao, and Taiwan returning to settle in their hometowns who need to use collective-owned land to build houses, the applications shall be handled in accordance with the relevant provisions of the *Law of the People's Republic of China on Land Administration*.

In 1995, when the former State Bureau of Land Administration issued the "Notice on Printing and Issuing 'Several Provisions on the Determination of Land Ownership and Use Rights'" (《关于印发确定土地所有权和使用权的若干规定》;document No. 26), it stated that if a homestead of a registered nonagricultural resident (including overseas Chinese) was rural, and the property rights to the house had not changed, then the construction land use rights of the collective land could be determined in accordance with the law. If the reconstruction of a house is not approved after its demolition, the land use rights revert to the collective.

In 1999, the "Notice of the General Office of the State Council on Strengthening Land Transfer Management and Prohibiting Speculative Land Dealing" (《国务院办公厅关于加强土地转让管理严禁炒卖土地的通知》; document no. 39) stipulated:

Rural people's residences shall not be sold to urban residents, and urban residents shall not be permitted to occupy and use rural collective land to build houses. The relevant departments shall not issue land use certificates and real estate certificates for illegally built and purchased residences.

Sixth, regulations on homesteads were put into place. From the *Regulations on the Construction of Villages and Towns* (《村镇建设条例》) of 1982 to the *Law of the People's Republic of China on Land Administration* of 1998, the regulatory arrangements for the regulation of homesteads and construction on them continued to take shape, until the Ministry of Land and Resources in 2004 issued the "Opinion on Strengthening Administration of Rural Homesteads" and formed the current control system. Two main features of this control system include the following.

The first main feature of this regulation system is that strict implementation of planning proceeds from strict controls on the scale of construction land in villages and towns. The planning and administration for rural homesteads is strengthened. Any occupation and use of agricultural land by rural homesteads should be included in the annual plan of governments at all levels. In the annual plan indicators on the use of agricultural land that the provinces (autonomous regions and municipalities) issue to counties (districts and cities), planning indicators for the occupation and use of agricultural land by rural homesteads may be added. The latter planning indicators should be linked to newly increased arable land area in rural construction land consolidation. After the county (city) land and resources management department inspects and confirms the newly increased arable land area, an equivalent amount of agricultural land within the annual planning indicators should be preferentially allocated for conversion into land indicated for rural housing construction.

The regulation system's second main feature is the improved homestead approval system and standardized approval process. Counties (cities) may, based on the planning quotas on the use of agricultural land for homesteads that the provinces (autonomous regions and municipalities) issue to counties (districts and cities) and on the actual need for rural villagers' housing construction, apply once per year at the beginning of the year to the province (autonomous region and municipality) or districted city and autonomous prefecture to handle the examination and approval procedures for the conversion of agricultural land, and after approval, the county (city) shall, in accordance with law, approve the provision of homesteads by household and clan. As for rural residents' use of vacant land, old homesteads, and unused land within their villages for residential construction, they must apply to the village (township), level by level; the county (city) approves these in batches, and then the villages (towns) distribute the homesteads by household and clan.[13]

2.2 The Main Features of the Rural Homestead System

First, the collective has ownership of the homestead. The rights of homesteads are clearly more complex and difficult to understand than those relating to agricultural contracted land. Whereas the collective's ownership of the homestead shows up when the collective is a contracting party, its ownership of the homestead is manifested more powerfully in other respects—mainly in that obtaining the use rights to a farming household's homestead must pass approval by the village committee. When the farming household is no longer using the homestead, the collective can take it back. When any improving or quotas' trading is conducted on the homestead, the main body concerned is also the village committee. Therefore, in the relationship between the ownership and the use of a homestead, the collective organization is not just something on the level of legal significance but has actual administrative and controlling power.

Second, farmers have the use rights on homesteads. By contrast, the farmer's rights to use the homestead are in fact concealed under collective ownership. As long as the farmer does not change ownership of the house, the homestead's existence is always ignored, and the farmer himself does not pay attention to it. As for the actual effectiveness of the rights, the *Real Right Law of the People's Republic of China* lists the homestead with usufructuary rights, and central policy documents have repeatedly emphasized protecting the farmers' usufructuary rights on their homesteads. However, the *Real Right Law of the People's Republic of China* also describes the empowerments of the usufructuary right as "the right to tenure and revenue" a property, but a homestead as a usufruct, in law, is empowered only with "the right to tenure and use"—it does not include "the right to revenue" the property. In this way, the property attributes of the homestead are greatly reduced, so that basically it has only a residential function.

Third, farmers have ownership of their houses. Since the CPC came to power, there has never been a change on the issue of farmers' property right of housing.

There has always been the insistence that rural houses are owned by the farmers, and their right to buy, sell, lease out, mortgage, pawn, and transfer their houses has been fully protected. Moreover, the government or collective organizations have always been prevented from infringing on property rights as they relate to farmer's houses. Therefore, in rural areas, farmers have formed the deeply ingrained idea that their houses are their homes, and that no matter what happens, they do not worry that their houses will be lost or pulled away from beneath them.

Fourth, the free obtaining of the homestead based on the membership. Since communization, it has always been the case that homesteads are acquired based on a farmer's membership status in the collective economic organization. In other words, only members of the collective economic organization are eligible to apply for and obtain homesteads, and nonmembers cannot. Since only the rights of occupation and use are granted in homesteads, there is no way for nonmembers to obtain them through transactions and transfers involving the homesteads. Because of the "one house per household" (*yihu yizhai* 一户一宅) rule, a farmer cannot reapply for and obtain a homestead after he transfers the house on it. This effectively blocks farmers from transferring the homestead use rights to nonmembers. Meanwhile, as long as farmer is a member of the collective organization, he can obtain a homestead free of charge. That is to say, under the current homestead system, the homestead is not a commodity, let alone a form of capital.

2.3 Problems in the Operation of the Homestead System

First, a large number of homesteads have entered the market. Even though the law does not extend the rights of leasing out, transferring, and trading homesteads, in actuality, putting farmers' homesteads on the market has become a general trend. Specifically, the trend appears in different ways in different regions. In the vast coastal regions, farmers have built multistory houses on their homesteads and then rent out the apartments, thus helping to house the large number of migrant workers who have left the farm to satisfy the labor demands of rapid industrialization. In the mixed urban-rural areas of large, medium, and small cities and towns, after the government has expropriated farmers' arable land, leaving them with the homesteads, the farmers use the homesteads to build housing that they rent out, thereby solving migrant workers' housing needs in the context of rapid urbanization. In the vast agricultural regions, some farmers who have left the farm for the city and found stable jobs and residences there privately transfer or lease out their homesteads to other farmers who need homesteads for building a house. The current situation of so many homesteads entering the market in different kinds of regions solves not only the problem that local people face in increasing income from their property during the industrialization and urbanization process, but also the housing issue for all the people entering the cities. It also lowers the costs of urbanization. However, this spontaneous entry of homesteads into the market is also in direct conflict with current laws.

Second, farmers' free obtaining of homesteads and the one house per household rule are difficult to implement. Free homesteads are considered a kind of welfare that the government gives to farmers, and they are a reflection of confidence in the system. However, as the coastal regions and mixed urban-rural areas undergo urbanization, there has been a substantial increase in the differential of income received for different grades of land, and as the supply of urban construction land tightens, the free obtaining of homesteads runs counter to the capitalization of land. In these regions, homesteads have not been freely distributed since the mid to late 1990s. Also, the one house per household rule is hard to manage in an urbanizing area. As the value of homesteads in these areas increases, farmers have made full use of them, adding more and more house units to the property, sometimes breaking outside of the areas of the original sites and sometimes breaking the rules that various provinces and cities have set for the floor space of houses on homesteads. Finally, in the traditional agricultural areas, free obtaining of homesteads does not make use of protected cultivated land. In these areas, homesteads are still freely distributed, but in recent years, as income from outside work increases, farmers return one after the other to their villages to build ever larger houses. If the original homestead is not large enough, they occupy and make use of their contracted land for the purpose. As a result, the amount of cultivated land in these agricultural areas, especially in the plains, is rapidly decreasing.

Third, disorderly expansion of homesteads is not conducive to the healthy development of cities. Because the way the homesteads are being used today is in serious conflict with the law, the government is basically absent from the actual management of the use of homesteads; planning and regulations on use cannot be implemented. In the absence of governmental regulatory control, the expansion of farmers' homesteads and the construction on them are becoming more disorderly, even spreading like vines. This has caused chaotic private construction in "urban villages" (*cheng zhong cun* 城中村), which are completely unplanned and lack basic infrastructure, public services, and governance organizations. Besides presenting a serious public security problem, they have formed a complete second skin over the city, which increases the cost of urban management and the difficulty of future renewal.

3. The Rural Collective Construction Land System: Features and Problems

3.1 Formation of the Rural Collective Construction Land System

In the early days of reform and opening up, the bulk of construction land was found in rural villages. On the one hand, as production responsibility was assigned to farming households (*bao chan dao hu*包产到户), farmers' incomes increased, so the first wave of farmers' housing construction emerged in the villages. On the other hand, after 1984, the rural surplus labor force became hugely prominent, so national policy vigorously encouraged farmers to develop village and township enterprises (*xiang zhen qi ye* 乡镇企业), and structural reforms were implemented to improve farmers' incomes and nonagricultural employment opportunities. For five consecutive years from 1981 to 1985, the growth of rural homesteads and construction land for commune- and brigade-run enterprises (*she dui qi ye* 社队企业) exceeded that of land designated for state construction; their growth was the leading cause of arable land being occupied. At first, if farmers needed a homestead, all they had to do was to submit a request to the production team leader or party branch secretary, and once approved, they could obtain a homestead for free. Production team enterprises could be set up merely by making an adjustment in members' contracted land. Commune- and brigade-run enterprises only needed to have an adjustment made on the production team's land or absorb some of the production team's labor and pay a small amount of compensation for above-ground things such as foliage, and the production team's land could be obtained. In 1981, the State Council issued the "Urgent Notice of the State Council on Prohibiting the Building of Houses in Rural Areas on Farmland" (《关于制止农村建房侵占耕地的紧急通知》; No. 57 [1981], State Council), and in 1982, the *Regulations on the Administration of Land Used for Housing Construction in Villages(*《村镇建房用地管理条例》) for formulated and promulgated. However, the *Regulations* did not at all contain this situation.

On the other hand, the Third Plenary Session of the 11th Central Committee (1978) considered the development of commune- and brigade-run enterprises (which, after the People's Communes were dissolved, became village and township enterprises) as an important move for revitalizing the rural economy. In the Suzhou area, during the three years from 1978 to 1980, the amount of cultivated land decreased by 210,000 *mu*, the state occupying 8.7%, commune- and brigade-led enterprises 34.7%, and farmers' homesteads 25.2%. In 1978, an estimated 2.355 million *mu* of land was occupied by village and township enterprises nationwide, and by 1985, the figure was estimated to be about 8.445 million *mu*—an increase by a factor of 3.6. In 1978–1979, construction in rural areas amounted to approximately 400 million square meters; in 1980, it was about 500 million square meters, and from 1981 to 1985, the average annual amount of new rural residential building exceeded 600 square meters. In 1978–1980, the amount of land occupied by rural housing construction in Jiangsu province was 3,600 hectares. According to 1985 statistics from the Ministry of Construction, housing construction by farmers nationwide occupied 60,300 hectares of cultivated land. In 1985, there was 1.979 million hectares of collective and township construction land nationwide. As for the occupation of cultivated land by village and township enterprises and houses built by farmers, this amounted to 77,000 hectares in 1982, 136,000 hectares in 1984, and 189,000 hectares in 1985. In 1982, rural collectives and individuals occupied 1.155 million *mu* of cultivated land nationwide for the purpose of residential construction; this figure reached 1.29 million *mu* in 1983, 2.04 million *mu* in 1984, and 2.835 million *mu* in 1985.

Responding to the severity of the problem of cultivated land being heavily occupied, the State in 1987 promulgated the *Law of the People's Republic of China on Land Administration*, which regulated collective construction land. Even so, the two channels for land acquisition—village and township enterprises' construction land and farmers' land for residential construction—have always remained open. As a result, collective construction land accounted for about half of all the land used for construction during the subsequent period. Let us take a look at the amount of land used each year for collective construction. In 1988, there was 690,000 *mu* of collective construction land; in 1989, 512,000 *mu*; in 1990, 576,000 *mu*; in 1991, 713,000 *mu*; in 1992, 935,000 *mu*; and in 1994, 713,000 *mu*. As for the area occupied by rural houses, in 1988 it was 821,000 *mu*; in 1989, 630,000 *mu*; in 1990, 555,000 *mu*; in 1991, 491,000 *mu*; in 1992, 462,000 *mu*; and in 1994, 431,000 *mu*.[14]

The promulgation and implementation of the new *Law of the People's Republic of China on Land Administration* in 1998 marked a turning point in the marketization of China's collective construction land. This law stipulated that in order for agricultural land to be converted into construction land, there had to be land expropriation; if land was needed for construction, state-owned land had to be used. Only one opening was left for rural collective construction land:

For the establishment of an enterprise using the land for construction determined by the overall planning for rural (township) land utilization by a rural collective economic organization or joint establishment of an enterprise with other units and individuals in the form of equity participation of land use right and joint operations . . . .

In fact, however, the channel by which agricultural land could legally be converted and brought into the collective construction land market was getting narrower and narrower. The reasons for this is, first, that the law recognized that the farmers' collectives had land on which they were running their own businesses or jointly operating enterprises, but after the 1998 reform of township enterprises, farmers' collectives no longer met the requirements for using land to operate their businesses under the law. Second, following the implementation of land-use scale controls and the approval management through the land-use quotas in 1997, the provincial governments naturally used the scarce quotas of land for the provincial capitals and other central cities, so that most county-level economic development seldom used quotas of land.

The first outcome of this situation was that the amount of collective construction land was greatly reduced. From 1998 to 2007, the amount of legal collective construction land truly used for construction was not at all comparable to that of the previous period. In many counties and cities, the construction land indicators that could be allotted to them amounted only to around 300–500 *mu*. This was not enough for county governments to use for attracting business investment, let alone for distribution to farmers' collectives, which would use it for nonagricultural construction projects.

The second outcome was that illegal land use proliferated. Since the construction land indicators could not be obtained, localities that wanted to develop their economy had little choice but to risk violating the law and just use the land first. Most of the collective construction land was also used for farmers' housing construction. Currently, the amount of rural residential land in China covers more than 170,000 square kilometers, and because there are no construction land indicators, let alone any approval, much of this construction land is illegal.

3.2 The Growth Effects of Collective Construction Land

First, the huge amount of collective construction land can provide space for the next round of growth. At present, in cities and their surrounding areas throughout China, the amount of collective construction land occupied by old villages, old factories, and old properties is huge. The experience of some pilot zones has informed us that the activation of this part of the land stock could break through the bottlenecks restricting China's economic development and free up new space for industrial development and urbanization. In Guangdong Province alone, there is more than 3.6 million mu of such land in stock, and that is equivalent to more than ten times the amount of new construction land. In Beijing, approximately 753 square kilometers of the central urban area is mixed urban-rural, within which there are 227 administrative villages (*xing zheng cun* 行政村), involving approximately 450 natural villages (*zi ran cun luo* 自然村落), that need to be rebuilt. These lands are close to densely populated and intensively industrialized areas, and the potential for differential land benefit is significant. However, due to restrictions on the transfer and mortgage of collective construction land, the value of this land is not fully manifest, which at the same time causes much of these land resources to be wasted. Based on Guangdong's experience with the "Renovation of the Three Olds,"[15] about 40% of the original collective construction land can be rebuilt and redeveloped. If we use this percentage as a basis for estimation, then if the properties on all of the more than 3 million *mu* of Guangdong Province's "Three Old" types of land can be rebuilt, the amount of land that could be saved would exceed one million *mu*. Meanwhile, the amount of new construction land in Guangdong Province by 2020 will be somewhat more than 2.6 million *mu*. Thus, the more than one million *mu* of land saved would be equivalent to 40% of the amount of Guangdong's new construction land in the next ten years.

Second, it increases land use intensity and doubles the benefit on land. Because of the lack of legal guarantees on usage of collective construction land in comparison with state-owned construction land, its main contractors are farmers acting on their own initiative, the development on it that takes place on it is, to a corresponding degree, disorderly, fragmented, and sporadic. Its industries are relatively low-end, and the low effectiveness of the land management is very noticeable. An improved collective construction land system would, on the one hand, stimulate rural residents' enthusiasm and, on the other hand, concentrate previously scattered land parcels for unified development, thus achieving a substantial jump in the output per unit area of land. For example, in Nanhai District (南海区) of Foshan (佛山), Guangdong, the factories or trade distribution centers built in the early stage of rural industrialization in the 1990s occupied large areas, but the plot ratios were very low—generally only 0.3—and the land utilization rate clearly tended to be low. In recent years, land use intensity has increased significantly by demolishing or renovating old buildings, linking the plots of different collective economic units, implementing unified planning, and introducing new projects. As land use intensity has increased, collective rental income has also increased. One such project was the rebuilding of the lumber market in Fengchi Village (凤池村), Dali Town (大沥镇), in Nanhai District. Since its transformation, the average monthly rent for a shop has risen from 12 yuan/m^2 to over 40 yuan/m2.

Third, it accelerates urban renewal and promotes industrial transformation. Over the past thirty-plus years of reform and opening up, as the industrial structure has been continuously upgraded, the economic form of collective construction land in cities and their vicinities has undergone earth-shaking change, while the collective economy has also transformed from "every agricultural enterprise is stable" to "every industry thrives" and again to "every business prospers." The implementation of urban renewal and industrial upgrading projects on collective land is entirely in line with the requirements of moving to the next economic growth stage, improving the quality of urbanization, and adjusting the industrial structure; it contributes to urbanization and the coordinated development of industry.

On the one hand, during the process of redeveloping on collective construction land, it may be possible to further uncover urban resource potential, enrich the social life of urban communities, readjust and optimize urban functional layout, achieve a more ideal integration of buildings and people, industries, culture, and environment, attract more people of talent, and provide a good environment for giving full play to the urban agglomerating effect, by changing building density, plot ratios, improving support facilities, and increasing green space. Pilot zones in Guangdong's Pearl River delta and the Beijing mixed urban-rural fringe have completely changed their local environments from their past low, shabby, chaotic condition, effectively beautifying them through uniform planning of their existing store of collective construction land.

On the other hand, it should be possible to eliminate backward production capacity and free up space with which to introduce higher-grade projects or enterprises, in line with local conditions, and to optimize and upgrade local industrial structure. For example, Zhenggezhuang (郑各庄), near Beijing, has established an industrial chain on its collective construction land, one led by construction but also integrating cultural tourism, sci-tech industry, education and scientific research, cultural creative industry, and real estate development. In recent years, new industries such as health care, geriatric care, headquarters economy, and industrial incubators have been added, thus completing an industrial transformation. Gaobeidian (高碑店), meanwhile, used collective construction land and homesteads to introduce a commercial street for classical furniture, traditional cultural residences, leisure tourism, and other industries attracting tourists. In the future it also plans to establish a Chinese medicine park and a street that promotes and sells international brand name goods, and to promote the continual upgrading of the industrial structure. In the Pearl River Delta region, through the "Renovation of the Three Olds" (*san jiu gai zao* 三旧改造), industry has been upgraded, a huge space for urban service industries has been created, and urban structure has been optimized.

Fourth, it expands investment channels and optimizes investment models. Urban renewal and industrial upgrading through the development of collective construction land are concrete manifestations of improved urbanization quality. In urban renewal, there are many investment opportunities in both the public infrastructure and social areas. From 2008 to the end of September 2010, Guangdong Province invested nearly 176.6 billion yuan in the "Renovation of the Three Olds," a figure that accounted for 5.03% of Guangdong's fixed asset investment in the same period. In the first three quarters of 2010, the percentage increased to 6.01%, showing an upward trend. This investment positively contributed to Guangdong's response to the impact of the international financial crisis and to the expansion of domestic demand. If Guangdong's more than 3 million *mu* of land could be used for the "Renovation of the Three Olds," one million *mu* could be freed up for redevelopment. Based on a rough estimate of investment density of 3 million yuan per *mu*, we can project a total investment scale of 3 trillion yuan, which is 1.6 times the scale of Guangdong's urban fixed assets investment in 2012.

Urban renewal on collective construction land will greatly optimize existing investment models, effectively alleviate the debt risks faced by city governments, and promote the balanced and sustainable development of cities. Since the original urban infrastructure investment was basically government-led, many local governments currently carry a heavy debt load. Moreover, under the original construction model, city governments often chose to expand the cities outward in order to balance costs and benefits and to reduce the cost of land expropriation and demolition. Furthermore, since they were constrained by the dual urban and rural systems, they often placed some of the large-scale infrastructure projects in the city centers, since the mixed urban-rural areas on the outskirts were more difficult to deal with, and this led to uneven urban development. For example, in Beijing, much of the infrastructure does not cover the mixed urban-rural areas. Currently, these areas are basically built and maintained by rural collectives and individuals, and the government only grants them a small amount of subsidy. If it were possible to use the stock of "live" collective construction land and permit the land to be collateralized, the rural collectives in these mixed urban-rural areas would have more resources or use more flexible means of increasing investment in some areas of infrastructure. The diversification of infrastructure investment entities could also alleviate the financial pressure (from being sole investors) on city governments and reduce the governments' expenditures. Meanwhile, the diversification would help complete the upgrading of urban public facilities in line with local conditions.

Fifth, it increases farmers' incomes and expands consumer demand. Industrialization and urbanization on collective construction land actually created a benefit distribution mechanism that differs from the current government-led land expropriation and urbanization model. The differential land-derived income created by the direct entry of rural collective construction land into nonagricultural use would be retained in the village, and because the farmers collectively retain collective ownership of the land, they would be able to enjoy the differential land-derived income for a very long time. Moreover, if their land is expropriated by the state, they would not be subject to the maximum 30 times compensation for the land in its original agriculture use. For example, after farmers' collective land in Nanhai District, Foshan, Guangdong, was incorporated into the city, their rental income on the collective land increased from 2.26 billion yuan in 2008 to 3.02 billion yuan in 2012; the two-level dividends of the village group increased from 1.68 billion yuan in 2008 to 2.67 billion yuan in 2012, so that the per capita income of the two levels of members increased from 2,347 yuan in 2008 to 3,516 yuan in 2013. Currently there is a total of 759,000 shareholders, which accounts for 99.4% of the rural residents, so the per capita net income of rural residents in Nanhai District increased from 11,158 yuan in 2008 to 16,673 yuan in 2012. Meanwhile the urban-to-rural income ratio narrowed from 2.33:1 in 2008 to 2.18:1 in 2012.[16]

3.3 Institutional Dilemmas in the Use of the Collective Construction Land System

First, the use of rural collective construction land exists in a state of extrajudiciality. Based on the result of the second land survey, the construction land area of Nanhai District is 797,500 *mu*, of which 565,500 *mu* is rural collective, accounting for 70.91% of the total construction land in the district. The total area of rural homesteads, meanwhile, is over 170,000 *mu*, close to 22% of the total. In Beijing's urban-rural fringe areas, there are 77 street villages and towns and 1,673 villages (communities), and more than 3 million *mu* of land are collective private-use plots. In Shenzhen City, there is 917.77 square kilometers of existing construction land, of which approximately 390 square kilometers belong to the economic organizations of the original rural collectives, and within these, only

95 square kilometers are being used legally. In Guangzhou City, 494.1 square kilometers are being used by "old cities and towns, old factories, and old villages," and of this, 356.7 square kilometers are classified as collective construction land, accounting for 72.2% of the total "land used by the Three Olds" (*sanjiu yongdi* 三旧用地).

Second, the lack of the collateralized financing function has greatly limited the output level of collective construction land and farmers' asset-derived income. The current *Law of the People's Republic of China on Land Administration* does not allow collateralization and financing on collective construction land, which limits the possibility of its capitalization. Because all investment projects on this land must be paid by own funds, without financial leveraging, this greatly raises the threshold for any industry that enters this collective land. In addition, the lack of mortgage financing for this land greatly reduces its market valuation and prevents any real improvements in its utilization efficiency. If we look at the situation in Nanhai District again, land development intensity is already close to 50%, and 71% of all construction land is collective. If mortgage financing were permitted for collective construction land, then, even if we calculate its mortgage value as half that of state-owned construction land, the financing scale of the entire Nanhai District would increase by 120%, the transfer price of collective construction land would double or triple at least, and production per unit of GDP would increase by a factor of 3–5 times.

However, at present only a small portion of the land use rights on collective construction land can be easily mortgaged, and the financial institutions engaged in that mortgage business are mainly local ones, like rural credit cooperatives and rural commercial banks. In the survey, the leaders of farmers' and collective organizations reflected that once mortgage financing is allowed for collective land, it would be possible to obtain formal financial support, and the industrialization and urbanization of that land could take a big step up. The face of these areas would fundamentally change. Farmers' collective asset equity could also be securitized, so that individual wealth would multiply and greatly increase farmer incomes.

Third, non-inclusion of construction land in urban-rural master planning has curbed its overall economic benefits. The current laws on land and urban planning stipulate that a village can be included in urban planning only after collective land has first been expropriated and converted into state-owned land, and only then can the various urban infrastructure facilities be extended to the village. As a result, there is no formal planning and preparation for the industries and factories on collective construction land. Shoddy private construction of housing in villages is widespread, and the fragmentation of the land has greatly reduced the scale effect of the land's use. Meanwhile, the various plans that the villages independently design are not recognized by the government. As a result, the municipal infrastructure costs are paid mainly by the village collective, which adds to the financial pressure on its organization.

In Beijing's urban-rural fringe, due to the lack of planning, the residents of many villages have blindly expanded the area of their homesteads and the height of their structures. They even occupy public land areas such as transportation routes, and as a result, the villages' layout is chaotic. One can say that because of the lack of land use planning (including industrial planning) in the collective construction-use areas of these villages, the land space that would otherwise have been available for replanning has disappeared, so that the future development space for these villages has been greatly reduced.

Fourth, the extensive use of collective construction land makes it difficult to upgrade the industrial structure. Even though the transfer and circulation of use rights to collective construction land began a long time ago, there are still no specific legal provisions on the registration of property rights, division and transfer of land, payment of taxes and fees, and so on. To circumvent legal and policy risks, the collective organizations can only adapt to situations—they private lease out the construction land, or they build factories that they lease out. This has resulted in the inefficient use of the land, used only for low-end industries, and scattered in its layout. It becomes difficult to achieve any scale, or build any superior-grade industrial park or industrial belt.

This extensive utilization mode of collective construction land, first of all, brings low economic benefits. Generally, a lease contract for a factory runs between 10 and 20 years, so the collective organizations can only obtain relatively fixed annual rents (even if there are increases, they are not high). Secondly, extensive utilization makes it difficult to upgrade the industrial structure. If an attempt is made to cultivate or introduce a higher-level industry (like a scientific or technical innovative industry) to collective construction land, the greatest change will be the need for the collective organization to bear the risks of screening the industry, as well as handling operations and management, but this would be very difficult for villagers, who are so used to holding fixed dividends, to accept in the short-term. In addition, if the original lease contract period is too long, and a new industrial tenant must be brought in to replace the original one before the end of the contracted lease period, the collective organization needs to pay a large sum for liquidation damages. This also affects people's willingness to upgrade the industries on their land.

Fifth, lagging reform of the governance structure of collective organizations has hindered efficient distribution of collective construction land. Collective economic property rights are naturally ambiguous, so that even after the collective land, the factories on it, and other assets are converted into equity, and the "assets become capital, and farmers become shareholders," the contradictions within the collective organization are still very prominent. This is especially apparent in the economically developed regions. Returning to Nanhai District again as an example, the present shareholders, wanting to protect their existing rights and interests, opposed the distribution of equity to more than ten controversial groups, such as "women who marry out" (*chujianü* 出嫁女) and retired servicemen, which led to a standoff between people representing different group interests in the village. During the 2009–2011 period, more than 80% of the agriculture-related public complaints were related to the issue of the distribution of collective economic returns.

In the long run, the collective economic organization governance structure that "divides only dividends, not debts" (*zhi fen hong, bu fen zhai* 只分红、不分债) is not conducive to industrial adjustment and upgrading on collective construction land, and it basically stymies any improvement to land use efficiency. For example, several economic unions (*jingji lianheshe* 经济联合社) in Nanhai District owe more than 5 billion yuan in historical debt, but the shareholders consider only their immediate interests and still demand the yearly dividends from the collective. They have never considered revitalizing collective construction land or optimizing the industrial structure in order to increase the collective's revenue, nor have they considered debt-sharing. In 2005, Zhou Village (Zhou cun 洲村) in Nanhai District established a new type of joint-equity cooperative (*gu fen he zuo she* 股份合作社). Through the villagers' shareholding, another company was set up to rent 200 *mu* of the rural collective economic organization's land. At that time, 2,020 of the 2,420 people in the village participated in the cooperative, and the annual income from rent was more than 15 million yuan. Subtracting 3 million, the profit was approximately 12 million yuan. But so far, the cooperative has not expanded the land that it rents out. One of the reasons is that the 400 villagers who did not originally become shareholders also wanted dividends, because all of this land was collective. Clearly, reform of the governance structure of collective economic organizations is lagging behind, and it greatly affects the flexibility and utility of collective construction land.

Notes

1 Du Runsheng杜润生, *Du Runsheng zishu: Zhongguo nongcun tizhi biange zhongda juece jishi* 《杜润生自述：中国农村体制变革重大决策纪实》 [*Du Runsheng's Own Report: Record of Major Decisions on Changes to China's Rural System*] (Beijing: People's Publishing House, 2005), P17–23.

2 See zhonghua renmin gongheguo guojia weiyuanhui bangongting 中华人民共和国国家农业委员会办公厅 (General Office of the State Agricultural Commission of the People's Republic of China), ed., Bongye Jiti Hua Zhongyao Wenjian Huibian 《农业集体化重要文件汇编》 [Compilation of important documents of agricultural collectivization] (Beijing: CPC Central Party School Press, 1981), P31.

3 Du Runsheng杜润生, *Du Runsheng wenji*, 《杜润生文集》[*Anthology of Du Runsheng]*（Taiyuan: Shanxi Economic Press）P722–743，P784–804.

4 See Du Runsheng's discussion of the agricultural collectivization movement in Ma Guochuan 马国川, *Gongheguo buzhang fangtanlu*《共和国部长访谈录》 [*Interviews with Ministers of the People's Republic*] (Shanghai: Joint Publishing Press, 2009), P4–45.

5 Zhou Qiren 周其仁, *Gaige de luoji* 《改革的逻辑》 [*The Logic of Reform*] (Beijing: China CITIC Press, 2013), P161–173.

6 Du Runsheng,杜润生 *Du Runsheng wenji*, 《杜润生文集》[*Anthology of Du Runsheng]*（Taiyuan: Shanxi Economic Press）.

7 *Translator's note*: The collective economic organization can set aside as much as 5 percent of the collective's land as mobile land (*jidong di* 机动地), which allows itself some flexibility.

8 Liu Shouying 刘守英and Gong James Kai-sing Kung 龚启圣, "Nongmin dui tudi chanquan de yiyuan ji qi dui xin zhengce de fanying 农民对土地产权的意愿及其对新政策的反应 [Farmers' Intentions on Land Property Rights and Their Response to the New Policies]," *Zhongguo nongcun guancha* 《中国农村观察》 [*China Rural Survey*] (1998.2): 20–27.

9 Zhao Yang 赵阳, *Gongyou yu siyong: Zhongguo nongdi chanquan zhidu de jingjixue fenxi*《共有与私用：中国农地产权制度的经济学分析》 [*Public Ownership and Private Use: An Economic Analysis of China's System of Rural Land Property Rights*] (Beijing: Sanlian Shudian, 2007), P84.

10 Tao Ran 陶然, Wang Hui 汪晖, and Huang Lu 黄璐, "Erlun chengbao hou de Zhongguo nongcun tudi xingzhengxing tiaozheng二轮承包后的中国农村土地行政性调整——典型事实、农民反应与政策含义 [*Administrative Adjustment of China's Rural Land after the Second Round of Contracting:Typical Facts, Farmers' Reactions, and Policy Implications*]," *Zhongguo zhidu bianqian de anli yanjiu* 《中国制度变迁的案例研究》 [*Case Studies in Chinese Institutional Change*] 10 (2010.00): 12–20.

11 Chen Xiaojun 陈小君, "Zhaijidi shiyong quan zhidu: guifan jiexi, shijian tiaozhan ji lifa huiying 宅基地使用权制度：规范解析、实践挑战及立法回应 [The House

site Use Rights System: Normative Analysis, Practical Challenges, and Legislative Responses]," *Guanli shijie* 《管理世界》 [*Management World*] (2010.10): 1–12.

12 Zhao Shufeng 赵树枫, "Shilun nongcun zhaijidi zhidu gaige 试论农村宅基地制度改革 [On the Reform of the Rural Homstead system]," *Nongcun gongzuo tongxun* 《农村工作通讯》 [*Newsletter About Work In Rural Areas*] (2006.10): 34–36.

13 Gao Shengping 高圣平 and Liu Shouying 刘守英, "Zhaijidi shiyongquan chushi qude zhidu yanjiu 宅基地使用权初始取得制度研究 [Study on the System for the Initial Acquisition of the Use Rights to House Sites]," *Zhongguo tudi kexue* 《中国土地科学》 [*China Land Science*] 21.2 (2007): 31–37.

14 Liu Shouying, *Zhongguo de eryuan diquanli zhidu yu tudi shichang canque* 《中国的二元土地权利制度与土地市场残缺》 [*China's Dual Land Rights Systems and the Broken Land Market*] (Beijing: Zhonggo sheke wenxian chubanshe, 2009).

15 It refers to "old cities and towns, old factories, and old villages."

16 Liu Shimian 刘世锦, Liu Shouying 刘守英, Xu Wei 许伟, and Shao Ting 邵挺, "Guangdong sheng Foshan shi Nanhai qu jiti jianshe yongdi rushi diaocha 广东省佛山市南海区集体建设用地入市调查[Survey on the Incorporation of Collective Construction land into the City in Nanhai District, Foshan City, Guangdong Province]," *Dongfang zaobao Shanghai jingji pinglun zhoukan* 《东方早报•上海经济评论周刊》 [*Oriental Morning Post Shanghai Economics Journal*] (December 10, 2013): 5.

3

China's Land Expropriation System: Features and Problems

1. Introduction

In any country or region, regardless of whether it uses a private or public ownership system, the land expropriation system is one of the most important systems used in converting land to nonagricultural use. Since different countries have different systems and levels of development (and use different terms to describe it, such as land acquisition, compulsory purchase, eminent domain, etc.), there are huge differences among them in terms of purpose and scope of land expropriation, the procedures involved, and the compensation, so the degree of impact it has on economic development also varies.

Over the past 30 and more years of reform and opening up, the Chinese economy has maintained a vigorous, high level of growth. The economy has maintained a growth rate of more than 10% per year. By 2011, China's industrialization rate had reached 46.8% and its urbanization rate had reached 51.3%. With these processes of industrialization and urbanization, a large quantity of land has been converted to nonagricultural uses. In comparison with economies that have undergone high-speed industrialization, land has played a very unique and important role in China's economic development. Since 1998 especially, China's high economic growth rates have been driven by the twin engines of high-speed industrialization and urbanization, but land functions as the ignitions of those engines. The local governments have established industrial parks through the low-cost expropriation of farmers' land and promoted high-speed industrialization through land lease agreements (letting the land out at cost or at zero cost). Using this land which they expropriate at low prices, they have a monopoly on the first-level land market, and promote the rental and mortgage financing services which they offer in their push for rapid urbanization.[1]

This chapter will sort out the formation of China's land expropriation system and its main features, analyze the role of the land expropriation system in China's economic growth, reveal the main problems in the current land expropriation system, and make policy recommendations for reforming the current land expropriation system.

2. The Formation and Main Features of China's Land Expropriation System

Land expropriation is the administrative act in which the state, based on the needs of the public interest, exercises its highest ownership right. The administrative organ that exercises the executive power on the state's behalf, following procedures set by law, obtains people's property by compulsory means and gives the property owners suitable compensation. There are four crucial elements in the exercise of land expropriation power by any government. First, in the planned public enterprise motivating the expropriation, or in the purpose of the expropriation, there must be a level of public interest sufficient to justify it; the matter of public interest must be a major and special one that undergoes a decision-making process. Second, the expropriation must have a legal basis and be carried out on the basis of established laws. Third, the expropriation must be in line with the principle of proportionality, that is, the place and the method for minimizing loss to the interested party should be selected. The scope of the expropriation must be limited to what is necessary and must not be indiscriminately expanded, in order to avoid excessive infringement on people's property rights. Fourth, fair and reasonable compensation must be given for any expropriation. Basically, land expropriation everywhere is based on these four elements, but the content of the various steps differs rather widely from country to country. In the course of China's economic development over the past 60-plus years, the uniqueness of the socialist system and, later on, the rapidness of industrialization and urbanization have meant that the objectives, content, and features of China's land expropriation system are also quite unique and constantly evolving.[2]

2.1 The Formation and Evolution of China's Land Expropriation System

We can divide the formation and evolution of China's land expropriation system into three periods, based on the relevant laws that were promulgated and implemented. During the first period, from 1953 to 1981, the state promulgated the 1953 *Measures on Land Expropriation for State-led Construction* and the 1954 *Constitution*, and in accordance with the *Measures*, guaranteed the supply of land for construction during the planned economy period. During the second period, from 1982 to 1997, the 1982 *Constitution* and the first *Law of the People's Republic of China on Land Administration* (1986) were promulgated; the latter law was in use until 1997. During the third period, from 1998 to the present, the revised *Law of the People's Republic of China on Land Administration* (1998) was introduced; this forms the basic framework of China's current land expropriation system. The implementation period of this law coincided with the period of China's rapid industrialization and urbanization starting in the late 1990s.

First period (1953–1981). After seizing power, the CPC promulgated the new China's first *Constitution* in 1954. At that time, because land was still mainly privately owned, this *Constitution* emphasized, "The state protects farmers' right of land ownership and other means of production in accordance with the law," and at the same time, it also for the first time clearly specified that the state would "purchase and expropriate urban and rural land for the needs of the public interest, and restore it to state ownership." These expressions provide us with three very important pieces of information: (1) land expropriation can be carried out only if there is a public interest need; (2) the object of land expropriation would include not only rural land but also urban land; and (3) the expropriation method would not be limited to confiscation but also include acquisition (i.e., with compensation).

With the start of the first Five Year Plan, large-scale construction projects were launched, so in 1953, China introduced the *Measures on Land Expropriation for State-led Construction*. For the most important element of land expropriation, the expression "public interest," which appears in the 1954 *Constitution*, does not appear; the relevant clause instead reads, "to adapt to the needs of national construction." As for the scope and uses of land expropriation, the law listed them, to wit, "for the building of factories, mines, and railways, for transportation, water conservancy, and national defense projects, for cultural, educational, and health facilities, municipal construction and other construction." As for compensation and resettlement, the law had several distinct features. First, resettlement was placed first. Before expropriation occurs, "... there must be proper resettlement for the production and living of the person from whom the land is expropriated. If the person from whom the land is expropriated cannot be resettled for a time, the expropriation should proceed after the resettlement has been done properly, or other land should be chosen for expropriating." To the present, this land expropriation law was the one that set the strictest and most explicit conditions for proceeding with expropriation. Second, the law stipulated that when the compensation was made, "adjustments to state-owned land and public land should be made as much as possible," and that it was only in cases where such land adjustments or their aftereffects would affect the production and lives of those whose lands were being expropriated that compensation or allowance would be paid. Third, the amount of land compensation was based on the total value of fixed production for the preceding two to four years; the housing compensation consisted of being given either an equivalent house, under the principle of guaranteeing that the original house occupants would have a house to live in, or compensation paid on a fair and reasonable basis; for attachments to the house, compensation would be paid on a fair and reasonable basis.

Second period (1982–1997). After the Third Plenary Session of the 11th CPC Central Committee of 1978 opened the curtain on China's reform and opening up, state governance was gradually put on a sounder legal track. In 1982, the first *Constitution* after reform and opening up was promulgated and implemented. It restored the 1954 *Constitution*'s "public interest" prerequisite for land expropriation, clearly stating, "The state may expropriate land in accordance with the law for the needs of the public interest." However, this *Constitution* for the first time stated, "Urban land belongs to state ownership. Rural and suburban land, homesteads, private retained plots, private reserved mountains, except where they are owned by the state in accordance with law, belong to collective ownership." As a result, the structure of two coexisting landownership systems, wherein "urban land is state-owned and rural land is collectively owned," has become one of the most important institutional arrangements in China's unique land expropriation system.

Based on the principles of the 1982 *Constitution*, China in 1986 introduced its first *Law of the People's Republic of China on Land Administration.* It had the following distinctive features.

First, adherence to the principle that land expropriation in the name of the law must be in the public interest. As mentioned above, the Constitution states quite clearly, "The state may expropriate land in accordance with the law for the needs of the public interest." Regrettably, the 1986 version of the *Law of the*

People's Republic of China on Land Administration did not clearly define which land uses would reflect the public interest, and it adopted a description of these uses that was even broader than the one in the 1953 version of the *Measures on Land Expropriation for State-led Construction*. That is, land could be expropriated wherever "the state proceeds with economic, cultural, and national defense construction and establishes social public utilities." It is worth noting that, in contrast to the 1982 *Constitution*, we find that the 1986 version of the *Law of the People's Republic of China on Land Administration* changed "may expropriate land" to "may expropriate collective land." We have not verified the reason for this, but this was undoubtedly the first time in the legal history of the new China that land expropriation was targeted at rural collectively owned land.

Second, the principle of urban-rural separation is followed. The *Law of the People's Republic of China on Land Administration* (1986), following the 1982 *Constitution*'s structure of two coexisting ownership systems, adopts the split governance of urban and rural land: "Urban land is state-owned, and rural and suburban land is collectively owned." Significantly, if we compare the *Law of the People's Republic of China on Land Administration* (1986) and the 1982 *Constitution* on the matter of the borders of state-owned land, the latter says, "Urban land (*chengshi tudi* 城市土地) is state-owned," while the former says, "land in urban districts (*chengshi shiqu tudi* 城市市区土地) is state-owned." Scholars who study urban land issues are well aware that China's expressions for cities vary quite widely, for example, "urban built-up areas" (*chengshi jiancheng qu* 城市建成区), "urban planning areas" (*chengshi guihua qu* 城市规划区), and "urban districts," which has resulted in a wide variance in the scope of what a "city" is, and hence the scope of governance in the divided urban-rural governance system also varies widely. Although we cannot determine why the *Law of the People's Republic of China on Land Administration* made this sort of change on the *Constitution*, it does link to the process of urbanization in China. Under the heading of "urban land owned by the state," and through the constant adjustment of city planning, a land nationalization trend, wherein rural land is drawn into city maps via land expropriation, becomes a reality. We must admit that the legislators of that time, by putting a limit on the content of the "city," were quite ingenious.

Third, limits to governmental examination and approval authority for expropriating land were defined. When formulating and implementing the *Law of the People's Republic of China on Land Administration*, China still had not adopted common international land planning and use regulations—it still followed the examination and approval processes of the planned economy era. To protect cultivated land, measures were taken to define the limits of governmental authority at all levels to review and approve the expropriation of land. Specifically, any expropriation of 1,000 or more *mu* of cultivated land and 2,000 or more *mu* of other land were to be approved by the State Council; an expropriation of 3 or fewer *mu* of cultivated land and 10 or fewer *mu* of other land were to be approved by the county-level people's government; and expropriations of land between those limits were to be approved by the people's government of the provincial city and autonomous prefecture, and if not passed, the matter would be have to be decided by the Standing Committee of the Provincial or Autonomous Regional People's Congress.

Fourth, the methods for determining compensation and resettlement based on the annual output of cultivated land were further clarified. Compared with the *Measures on Land Expropriation for State-led Construction* (1953) (《国家建设征用土地办法》), the *Law of the People's Republic of China on Land Administration (1986)* first increased the compensation many times over, so that it became three to six times that of the average annual output value for the three years prior to the expropriation of the cultivated land. Secondly, the resettlement allowance would also be multiplied, based on the output value of the cultivated land. That is:

The resettlement allowance of the expropriated cultivated land is to be calculated based on the number of people to be resettled. The resettlement allowance standard for each agricultural population that needs to be resettled shall be three to four times the average annual output value per *mu* for the three years prior to the expropriation of the cultivated land.

The law also specified that the sum of the land compensation fee and the resettlement allowance could not exceed 20 times the average annual output value of the three years prior to the expropriation of the land. As for the payment of compensation, only the compensation for attachments and young crops is paid to the land-expropriated person himself. The compensation for the land and the resettlement allowance was to be paid to the land-expropriated unit, and this was to be retained and used to develop production, arrange for the employment of surplus labor created by the land expropriation, and pay living allowance for unemployable persons.

Fifth, arrangements for the re-employment of land-expropriated farmers were implemented. Besides the land and resettlement compensation, the *Law of the People's Republic of China on Land Administration (1986)* also included measures for the employment and change of status of land-expropriated farmers. First, it arranged for some land-expropriated farmers to enter the employment of the unit taking use of the original land or the collective or enterprise belonging to the whole people taking possession of the original land. Second, if all of the land of the land-expropriated unit was expropriated, then the household registrations of unit members were converted from agricultural to nonagricultural. In the early days of reform, since there were obvious advantages of being in a "whole people's" or collective ownership unit, re-employment and resettlement were still more attractive to land-expropriated farmers. However, with the development of multiple ownership economies, especially the nonpublic economy and foreign-invested economy, the vast majority of collectives and units belonging to the whole people were at a competitive disadvantage, and when the land-expropriated farmers got placed in these units, they usually became the first targets in layoffs. As a result, this resettlement method became less and less attractive to land-expropriated farmers.

Third period (1998 to the present). As China's industrialization and urbanization have accelerated since the end of the 1990s, the nonagriculturalization of land has also greatly accelerated, so that the threats to cultivated land and national food security are becoming increasingly serious. For this purpose, when China revised the *Law of the People's Republic of China on Land Administration* in 1998, one of the more important changes was to reform the methods of land expropriation.

The 1998 version of the *Law of the People's Republic of China on Land Administration* retains and improves some of the basic principles of the 1986 version. For example, it still follows the principle, in the legal sense, of land expropriation in the public interest. The principle of separation of urban and rural is not only continued, but further strengthened. As for compensation and resettlement from land expropriation, the original principle of compensation based on original use, but with the standards raised. Specifically, the amount of land compensation was raised to six to ten times the average annual output value of the cultivated land in the three years prior to its expropriation; the amount of resettlement subsidy was raised to four to six times the average annual output value of the cultivated land in the three years prior to its expropriation. Moreover, the sum of the two items above was not to exceed 30 times the average annual output value of the land in the three years prior to its expropriation. While the review and approval process still followed the approval management approach, the power of review and approval was further limited. That is, in the expropriating of basic farmland, if the amount of cultivated land in excess of the basic farmland exceeds 35 hectares, and the amount of other land exceeds 70 hectares, it would have to be approved by the State Council.

In addition, the 1998 version of the *Law of the People's Republic of China on Land Administration* made major changes in the following three areas, which have since had great impacts on land administration, land transfer, and land income distribution in China.

The first was to establish a "land use regulation system." The state prepares a master plan for land use, stipulates land use, and divides the land into agricultural land, construction land, and unused land. The conversion of agricultural land into construction land is severely restricted, the total amount of construction land is controlled, and special protections are implemented for cultivated land. Units and individuals using the land must do so strictly in accordance with the purposes determined by the master planning of land use, because, within the land use regulation system, the land use master plan is legally binding. Through tiered review and approval of the land use master plan, the amount of protected cultivated land, the scale of construction land, and the amount of land used in each category is controlled from above. The master planning of land use constrains scale of construction land in the urban master plans and the village and town plans. Controls on total land use and annual land use are implemented through the annual planning of land use.

The second was to stipulate the objects of using and means of obtaining construction land. First, any unit or individual that is building and needs land to do so must apply for the use of state-owned land in accordance with the law; this includes both state-owned land and land that the state has expropriated from farmers' collectives. Second, if construction on occupied land involves the conversion of agricultural land into construction land, it is necessary to go through the approval procedures for such a conversion. To expropriate agricultural land, one needs to handle the conversion of agricultural land first.

The third was to implement a system for the paid use of state-owned land. In 1998, the *Constitution* was amended, from "No organization or individual may misappropriate, buy, sell, lease out, or in any other form illegally transfer land," to, "No organization or individual may misappropriate, buy, sell, or in any other form illegally transfer land. The use rights of land may be transferred in accordance with the law." The *Law of the People's Republic of China on Land Administration (1998)* also clearly stipulated, "The use rights of land may be transferred in accordance with the law," and, "The state implements a system of paid use of state-owned land in accordance with the law." In its capacity as landowner, the state grants land use rights to land users within a certain period of time through agreement, bidding, and auction. The land user then pays the land use rights transfer fee according to the transfer contract.

2.2 Characteristics of the Current Land Expropriation System, with Evaluations

By constantly evolving, China's land expropriation system has formed an institutional arrangement with its own distinctive characteristics. We shall summarize them as follows:

(1) *A generalization of land expropriation under the principle of the public interest.* In the past 60-plus years, apart from the 1953 *Measures on Land Expropriation for State-led Construction*(《国家建设征用土地办法》), the various versions of the *Constitution* and their amendments, and land administration laws and their revisions, have all clearly stated that land expropriation must be to fulfill the demands of the public interest. However, so far, no law has given the meaning of the public interest to which they refer, and no law has clearly defined the range of purposes of land expropriation for the public interest. Not only that, but China's regulations on land acquired by means of expropriation are becoming more and more broad. The 1953 *Measures on Land Expropriation for State-led Construction*(《国家建设征用土地办法》) was more detailed about listing the uses of expropriated land: ". . . for the building of factories, mines, and railways, for transportation, water conservancy, and national defense projects, for cultural, educational, and health facilities, municipal construction and other construction." Judging from the kinds of projects listed, except for the building of factories and mines, the other items do basically have a "public interest" nature. By the time of the 1986 version of the *Law of the People's Republic of China on Land Administration*, however, the kinds of projects for which the state could expropriate land had become "economic, cultural, and national defense construction and for the establishment of social public enterprises." This is obviously broader in terms of the range of purposes than that stipulated in the 1953 *Measures on Land Expropriation for State-led Construction* (《国家建设征用土地办法》). When the 1998 version of the *Law of the People's Republic of China on Land Administration* was introduced, not even the general range of purposes for land expropriation was listed—the relevant clause had become, "Any unit engaged in construction must use state-owned land." The land available for construction, in addition to land that was originally state-owned, was collective land expropriated from farmers. That is to say, since the enactment and implementation

of the 1998 version of the *Law of the People's Republic of China on Land Administration*, what has in fact happened is that all construction must now use state-owned land that has gone through the process of expropriation, and this trend is becoming more and more obvious as industrialization and urbanization progress. From this, we can see that even though China has always stated in its laws that land can be expropriated to meet the needs of the public interest, no law thus far has defined what the public interest means or has set any clear boundary to the range of purposes for land to be expropriated.

This is quite at odds with the practice of the major market economies, which restrict land expropriation and reduce the erosion of land property rights by defining the public interest. In China's laws and regulations and in its actual operation of land expropriation, any engagement in economic construction is virtually equivalent to the public interest. It is precisely because of the generalization of the public interest in land expropriation that it has become convenient for the government and land users to obtain farmers' collective land by expropriation. During the planned economy era, since economic construction was always for the state's interest, drawing an equivalence between economic construction and the public interest could still be justifiable. However, as China's socialist market economy has improved, land-using entities and interest structures have increasingly diversified—and yet all construction is equated with the public interest, and the expropriation of land on a grand scale has become the usual practice. This has opened the door of convenience for public power to legally infringe upon private rights; this is contrary to jurisprudence, nor does it conform to social fairness and justice or the basic principles of the market economy.

(2) *The separate administration of urban and rural under two ownership systems*. The 1982 *Constitution* divides the land into urban state-owned land and rural collectively owned land. This is the most important institutional arrangement that distinguishes China's land system from that of other countries and regions in the world. To implement the landownership structure of this constitutional urban-rural division, two different sets of legal governance have been adopted for China's urban and rural land. Rural land is regulated by the

Law of the People's Republic of China on the Contracting of Rural Land, so the use, derivation of income from, and circulation of farmers'

collective agricultural land are subject to that law. Meanwhile, land transfer and urban state-owned law are regulated by the *Law of the People's Republic of China Land Administration*, so the conversion of agricultural land to nonagricultural use and the use, derivation, and transfer of use of that land after it has been transferred are protected by that law. From this, two completely different systems of rights appertaining to urban and rural land have been formed. Rural land users enjoy the rights to use, derive income from, and transfer use rights to agricultural land. As soon as agricultural land is converted to nonagricultural, the original rural landowner (the collective) not only loses its ownership but also loses its rights to use, derive income from, transfer, and develop that land. Farmers' right to use agricultural land in a nonagricultural way is restricted only by a clause in Article 43 of the *Land Administration Law of the People's Republic of China*—they can only "build township or town enterprises or build houses for villagers, after having lawfully obtained approval of using the land owned by farmers' collectives, or build public utilities or public welfare undertakings of a township (town) or village, after having lawfully obtained approval of using the land owned by farmers' collectives."

(3) *Conversion of agricultural land under many layers of regulation.* Based on the experience land management in countries and regions around the world, the conversion of agricultural land to nonagricultural land must be subject to planning and use regulations. However, land transfer in China is much more complicated, since it is subject to three layers of controls.

The first layer is planning and use regulation. When the *Law of the People's Republic of China on Land Administration* was revised in 1998, China studied the land management methods of countries with market economies, then implemented a land-use regulation system through a master planning of land use. At the same time, the construction and use of land after its conversion to nonagricultural use became subject to the cities' overall urban planning. Judging from the functional divisions and linkage relationship between the two plans, the function of the master planning of land use is to restrict the conversion of agricultural land to construction land and protect cultivated land; it is used to frame the boundaries of land use and make use regulation effective. The function of a city's overall urban plan is to make sure that use of every parcel of land within its borders is defined and implemented. However, since

China is rapidly industrializing and urbanizing, cities are rapidly growing in scale and number, and their boundaries are constantly expanding outward. It has become the norm now for city governments to expand room for development by adjusting their administrative divisions and revising their urban plans. Due to the cities' continuous expansion and shifting boundaries outward, the power of the land use master plan to restrict cities' border-shifting has weakened, and it has become difficult to resist urban sprawl and the conversion of agricultural land to nonagricultural land.

The second layer is landownership control. According to the principle of separate administration of urban and rural land, rural land belongs to the collective ownership of farmers, while urban land belongs to the state. Rural land must be expropriated in order for it to be changed into construction land, and if it is incorporated into urban territory, then that land must be changed from its original rural collective ownership to state ownership. This creates a large difference between China and the major market economies in terms of land conversion: in the latter, as long as planning and use regulations are followed, then landowners not only can use the land for nonagricultural construction but can also dispose of and transfer the land. They can also earn the differential income from the land's conversion to nonagricultural use. However, in China, due to the imposition of controls on ownership, a huge gap has been set between different owners in terms of land use rights and the distribution of profit from land. For farmers' collectives, even if their land belongs to a nonagricultural construction zone in the plans and use regulations, they have no right to use it or derive income from it. Instead, they must transfer it to the scope of urban planning and transfer it to state ownership, and then the local government assigns it to other users. The original farmers then collectively lose their rights to use, derive income from, and develop the land.

The third layer consists of annual land plan quotas and approval management. Due to the failure of land use regulation and the softness of the constraints on urban land use planning in urban expansion, plus a low-cost land expropriation system that encourages local governments to expropriate more, tenure more, and convert more rural land into urban land, the central government has had to strengthen the annual indicators and the management of the application-and-approval process for converting land to construction-use, in order to restrict local governments from using land use planning to expand cities. However, since the management by plan indicators and approval systems both carry a strong planning tinge, this top-down allocation of indicators not only has difficulty adapting to the needs of economic development, but also has difficulty really restraining local governments' use of land. The allocation efficiency is also low, and it creates rent-seeking problems.

(4) *Compensation according to the original use of land in land expropriation and resettlement.* As countries and regions with market economies develop, the volume of land expropriation will continue to decline, and the compensation for land expropriation will continue to be oriented toward marketization and property compensation. In China, since there is no land property market, landownership has no value. Therefore, when the government expropriates farmers' land, it has never taken the approach of compensating for the land property value. Even so, when expropriating land, the Chinese government has constantly faced problems due to the increasing land value and increasing rights awareness among farmers. To cope with rising farmers' demands, the Chinese government has mainly adopted the approach of occasionally multiplying the amount of compensation based on the land's agricultural output value and continuously improving the compensation standards. Compensation for land in the 1953 *Measures on Land Expropriation for State-led Construction* was based on the total output for the previous two to four years; in the 1986 version of the *Law of the People's Republic of China on Land Administration*, compensation was based on the annual output value of cultivated land and calculated at three to six times the average annual output value for the previous three years. By the time that the 1998 version of the *Law of the People's Republic of China on Land Administration* was enacted, the government still firmly adhered to the principle of calculating the compensation based on the land's original use, but the compensation standard was increased to six to ten times the average annual output value of the land for the three years prior to its expropriation.

Since compensation based on land property has not been implemented, China must use resettlement and employment as remedies in land expropriation. In 1953, the procedure was to put resettlement before expropriation—that is, without resettlement, the land could not be expropriated. In the 1986 version of the *Law of the People's Republic of China on Land Administration*, monetary settlement for the farming population was set at a multiple of the output value of the land, and the resettlement subsidy standard for each agricultural population to be resettled was set at two to three times the average output value per *mu* for the three years prior to the expropriation. The 1998 version of the *Law of the People's Republic of China on Land Administration* raised the resettlement standard, so that the resettlement subsidy standard for each agricultural population to be resettled was set at four to six times the average output value per *mu* for the three years prior to the expropriation. Clearly, this kind of labor settlement based on multiples of the farmland's output value cannot guarantee the long-term livelihood of farmers who have lost their land.

Besides resettlement, for employment and transfer of household registration are used in land expropriation. During the planned economy era, farmers who lost their land due to land expropriation were resettled by the state. First they went from "agricultural to nonagricultural," that is, their agricultural household registries were converted into urban ones. The second thing done was to arrange for the agricultural laborers to be employed in an urban enterprise or institutional unit and become employees. Under the dual system of that time, this made a fundamental change in the farmer's status. He and his children could enjoy all the employment and benefits of people in the city, and the social conflicts caused by land expropriation were not as sharp or widespread. Many farmers welcomed, or even hoped for, the state's expropriation of their land, since they considered it a chance to transform their own destinies. During and after the 1990s, however, China gradually established a socialist market economic system, and state-owned enterprises also gradually restructured. The nonpublic economy developed to the point that it accounted for half the economy. Employment and resettlement lost their meaning as a guarantee to farmers who had lost their land. For farmers who lose land through land expropriation, the state no longer resettles them—it has switched to increasing the standard by which the resettlement compensation is calculated, the compensation is entirely monetary, and the resettlement subsidy is also estimated based on multiples of the annual agricultural output value of the land. But in fact, once a farmer completely loses his land, it is impossible to farm, and if he is employed outside of agriculture, there is a long or short period of time when the cost of living for him greatly increases. It is difficult to maintain a normal life with the compensation for agricultural labor, so the standard of living for farmers whose land has been expropriated actually decreases. As a result, social conflicts caused by land expropriation have gradually become conspicuous. As these social conflicts have intensified, some localities have explored and implemented guarantees to maintain a minimum standard of living for land-expropriated farmers. The Ministry of Land and Resources and Ministry of Human Resources and Social Security have jointly issued a document asking that local governments establish a social security system for land-expropriated farmers, but to date, the proportion of the localities that have done this is not high. Even so, since this system provides only a low level of protection, it can hardly be considered commensurate with the land development rights that the farmers have lost.[3]

(5) *Land appreciation income returns to the government.* The land system arrangements described above make China quite unique in that regard. First, rural land belongs to the collective ownership of farmers, and farmers collectively possess the rights to use, derive income from, and transfer farmland, and when agricultural land is converted to nonagricultural use, the farmers' rights on the land are lost as soon as they receive compensation based on a multiple of its annual output value during its original use. Meanwhile, urban land is state-owned, and local governments enjoy the rights of disposing of, assigning, and deriving income from construction land. Second, the government has a monopoly on land, and local governments become the only legal channel for converting agricultural land into construction land. With one hand they expropriate land from farmers, and with the other hand they exclusively assign use of the construction land under their own name. Third, because the government is deeply involved in the nonmarket allocation of land, the implementation of the application and approval system for the occupation of agricultural land, the implementation of the land use planning controls on construction land, the marking off of construction land, and the negotiations on assigning the use of that land, the government controls the sales[4] and pricing of commercial land (*jingyingxing yongdi* 经营性用地). Fourth, governmental administration and business administration are united. The local government is both a land manager and land operator. In this uniquely structured land system, there are two core institutional arrangements: the compulsory, low-cost land expropriation system and the paid use system on construction land under the government monopoly. Under this system, land expropriation has become the

only conduit for converting agricultural land into nonagricultural land, and state-owned land has become the only land that can be used legally for nonagricultural construction. As industrialization and urbanization progress, farmers lose their landownership and their right to develop the land after the receive compensation based on multiples of the output value of the land in its original use. The government thus becomes the main earner in land value-added income during the industrialization and urbanization process.

3. The Land Expropriation System and Economic Development

3.1 Using Compulsory, Low-Cost Land Expropriation to Drive Economic Growth

Over the past 30-plus years of reform and opening up, the Chinese economy has maintained vigorous, high-speed growth. The economic growth rate and industrial added-value growth rate have remained in the double digits, and more than half of China's population now lives in cities. China has become the world's second largest economy and the world's largest exporter. In comparison with the features of high-speed industrialization and urbanization phases of other economies, land in China's last round of economic development has played a very unique and important role. The unique land system that China implemented during this period has been one of the secrets to its achievement of such high economic growth.

China's high rate of economic growth was supported by high investment and export levels and regional competition led by the government, and the land system was an important factor ensuring the success of this growth model. First, during the last phase of China's high-speed development, due to insufficient domestic demand and lack of innovation, economic growth was driven mainly by high investment and export levels, and on reliance on increased investment and expanding scale. This approach to development can only be supported through high consumption of resources. In factor input, because of the huge reservoir of rural surplus labor, the supply of low-cost labor was a major advantage. Land was another important input factor. Even though China's land resources are scarce due to population pressure, and lack of land is a basic national condition, the implementation of a dual land system of urban-rural inequality, a low-cost compulsory land expropriation system, and a nonmarket system for supplying land together enabled China to avoid the high land costs that would otherwise have affected economic growth. China could instead rely on an ample, guaranteed land supply to support economic growth. Secondly, in government-led regional economic competition, taking the initiative to attract investment and creating a suitable policy environment were the two most important magic weapons for local governments to beat the competition. Since land was the most valuable resource that local governments could control, using that land became their main investment promotion technique. Their main tools for keeping business enterprises local were incentives on land revenue and land tax. The ample supply of land and high consumption of resources ensured high levels of investment; a low-cost land supply kept land prices low, which guaranteed a high level of exports; and using land to attract business investment guaranteed rapid industrialization and urbanization. All of these promoted high rates of growth in local economies.

In the last round of economic growth, the ample supply of land played a fundamental role. From 1996 to 2010, the total area of construction land nation-wide increased by 74.1 million *mu* from 437.71 million *mu* to 511.81 million *mu*—a 16.9% increase, with an average annual increase of 5.293 million *mu*. Since 2006 especially, the scale of China's construction land has grown even more swiftly. In 2010, the total land supply nation-wide reached 6.45 million *mu*, an increase of 83.8% over the 2006 figure. During that period, the total amount of new construction land increased by more than 7.14 million *mu* per year, so that by 2010, the total area of new construction land nation-wide went as high as 10.68 million *mu*. Not only has the total amount of construction land increased significantly, but the supply of various types of land has also increased significantly (see Table 3.1 and Figure 3.1).

It can be seen from Table 3.2 that land supply has a high correlation with economic growth. From 2003 to 2010, during years with a large supply of land, the GDP growth rate was also higher, and during years when the supply of land decreased, the GDP growth rate was also relatively low (see Table 3.2 and Figure 3.2).

Most of the newly added construction land was acquired through land expropriation. From 2005 to 2010, the total area of land approved for expropriation averaged more than 4 million *mu* per annum, and it accounted for 78% of the newly added construction land. The supply of expropriated land also showed a growth trend. It is evident that even though China has implemented the most stringent farmland protection system in the world, it has not thereby reduced the supply of land. In particular, the guarantee of expropriated land for construction use has become an important factor in promoting economic growth (see Table 3.3 and Figure 3.3).

Table 3.1: Growth of various types of land use, 2003–2010 (unit: hectare).

Year	Industrial, mining, and storage land	Commercial use land	Total residential land	Public administration and public service land	Special land	Transportation land	Water and water conservancy land
2003	116,379.4	53,676.46	63,751.63	35,145.51	3,552.02	10,398.81	3,532.8
2004	106,756.7	43,831.1	59,689.78	27,189.5	2,697.05	13,456.98	4,298.58
2005	109,813.8	29,591.38	55,144.85	23,481.34	4,123.94	14,779.49	7,334.64
2006	154,635.3	32,124.51	65,153.68	29,582.6	2,239.51	18,135.89	4,934.41
2007	141,723.4	57,751.08	80,174.84	33,248.17	1,945.48	20,258.82	6,872.11
2008	92,918.09	26,532.02	62,030.09	26,573.24	2,623.58	17,004.19	6,503.5
2009	141,486.5	27,570.86	81,548.17	43,763.85	1,761.48	45,613.76	19,436.83
2010	153,977.6	38,905.15	115,272.5	52,945.57	2,522.37	48,795.49	19,490.72

Source: *Zhongguo guotu ziyuan tongji nianjian* 《中国国土资源统计年鉴》 [*China Land and Resources Statistical Yearbook*](successive years).

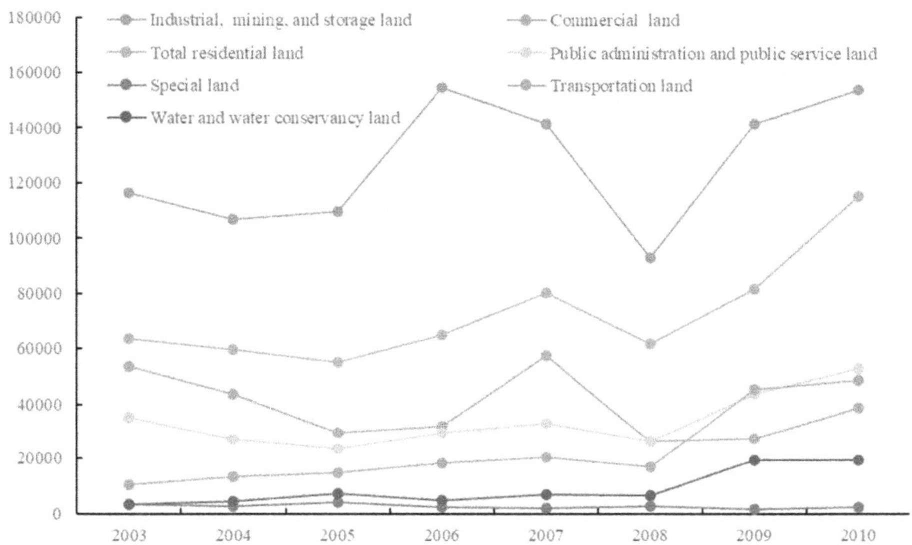

Figure 3.1: Growth of various types of land, 2003–2010 (unit: hectare).

Table 3.2: Total land supply and GDP growth rate, 2003–2010.

Year	Total land supply (hectares)	GDP growth rate (%)
2003	286,436.66	10.6
2004	257,919.7	10.4
2005	244,269.47	10.7
2006	306,805.89	13.3
2007	341,973.95	14.6
2008	234,184.68	9.8
2009	361,648.75	8.9
2010	432,561.42	10.8

Sources: *Zhongguo guotu ziyuan tongji nianjian* [*China Land and Resources Statistical Yearbook*] and *Zhongguo tongji nianjian* 《中国统计年鉴》[*China Statistical Yearbook*].

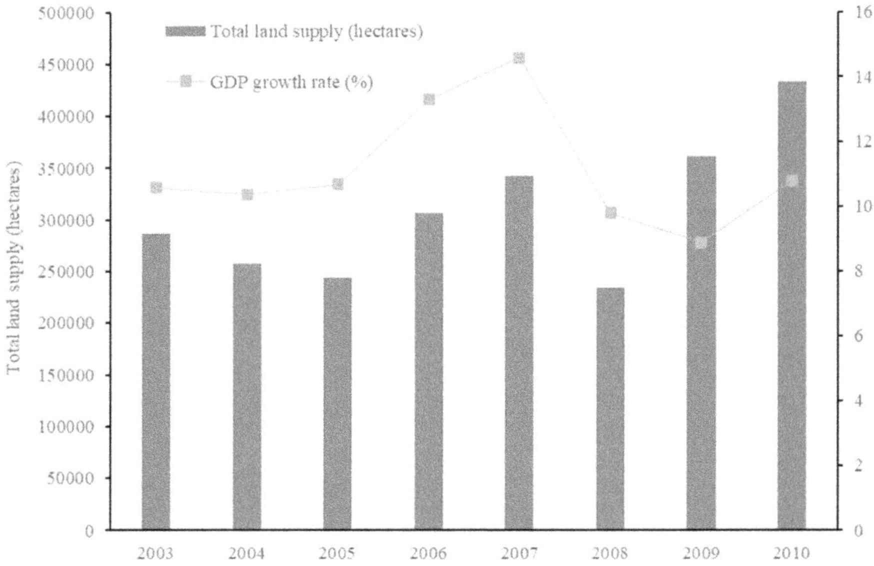

Figure 3.2: Total land supply and GDP growth rate, 2003–2010.

Table 3.3: Land expropriation nationwide, 2005–2010 (unit: 10,000 *mu*).

Year	Land area approved for expropriation nationwide	Agricultural land area approved for expropriation	Cultivated land area approved for expropriation
2005	445.4	350.1	242.0
2006	512.5	380.7	254.6
2007	466.4	334.7	222.4
2008	456.0	334.8	223.7
2009	676.5	526.8	325.1
2010	688.9	517.8	343.0

Source: *Zhongguo guotu ziyuan tongji nianjian* [*China Land and Resources Statistical Yearbook*] (successive years).

3.2 Using Low-Cost Land to Promote High-Speed Industrialization

In the last round of high economic growth, the rapid advancement of industrialization was a dominant force. From 1979 to 2011, China's industrial added-value grew at an average annual rate of 10%, and the industrialization rate reached 47% by 2011. Since the mid-1990s especially, thanks to an export-oriented industrialization strategy, China became the "world's factory" in just over a decade.

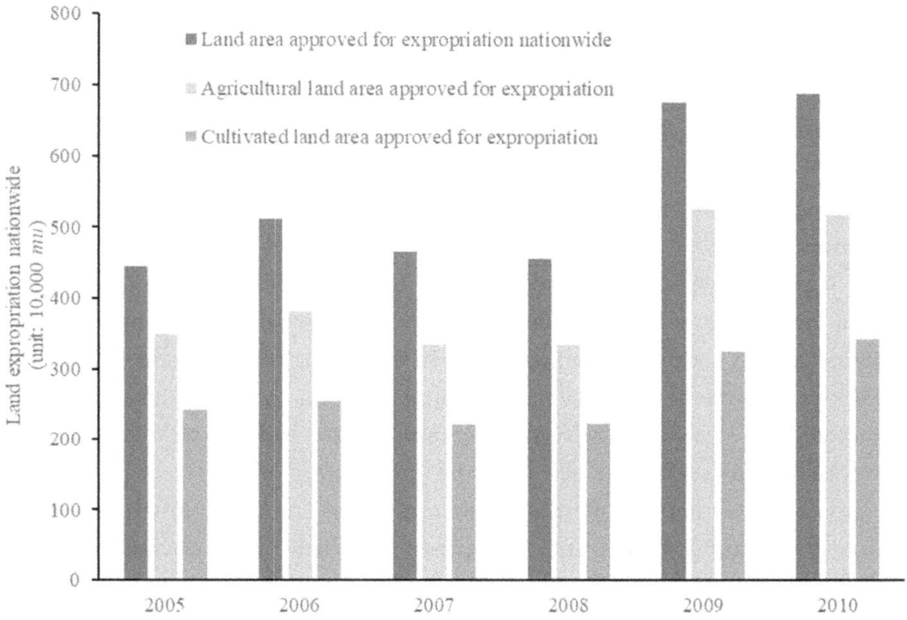

Figure 3.3: Land expropriation nationwide (unit: 10,000 *mu*).

China's system of compulsory low-price land expropriation and the exclusivity of the government's allocation of land resources are important factors driving its high industrial growth. On the one hand, when local governments use the land expropriation powers granted to them by the law and give land compensation and resettlement subsidy to the land-expropriated farmers based on the output value of the land as originally used, that is only equivalent to six to ten years of the income derived from the land. On the other hand, the local governments control the primary market for land, and they hold the authority to supply construction land in their own hands, thereby ensuring the supply of land for industrial construction at nonmarket prices. This unique land system arrangement not only enables China's industrial development to avoid the land scarcity problem but also makes the land a tool for attracting investment and promoting industrial development. Under this institutional arrangement, local governments take advantage of their convenient grasp of land resources to attract investment, using very large concessions on land prices in exchange for relatively scarce capital, technology, and the prospect of future tax revenue. The quick combination of low-cost land and low-cost labor with foreign capital and technology became an important factor in speeding up industrialization. In some places, local governments even supply land below cost or at no cost in order to recruit or retain business enterprises with tax potential, and in other places, local governments have formulated policies that set price ceilings for industrial land use agreements. The competition among regions to attract investment evolved directly into a vicious cycle, with each region lowering its land prices, and behind it was a financial competition among their local governments, resulting in long-term low prices for industrial land. Thus a unique phenomenon of the Chinese land market has been the scarcer the land resources, the lower the land price.

It is precisely within this unique land system arrangement that China allocates a large amount of scarce land for industrial use, thus ensuring the rapid development of industry. Compared with the major market economies, such quantities of industrial land are unusually high, and they are higher than the industrial land levels that these countries and regions had during their period of most rapid industrialization (see Table 3.4 and Figure 3.4).

It is precisely because of this unique industrial land system, where land in China is becoming scarcer and land prices more and more obvious, that the amount of industrial land has been able to stay at a low level for a long time. From the changes in land prices in 21 cities across the country, we can see that from 2000 to 2010, the national comprehensive land price level, commercial land price level, and residential land price level increased by 297.08%, 308.76%, and 528.10%, respectively, but the industrial land price level increased only by 71.18% (see Figure 3.5).

Table 3.4: Industrial, mining, and warehousing land as a share of total land supply, 2003–2010.

Year	Total land supply (hectares)	Industrial, mining, and warehousing land (hectares)	Percentage (%)
2003	286,436.7	116,379.4	40.6
2004	257,919.7	106,756.7	41.4
2005	244,269.5	109,813.8	45.0
2006	306,805.9	154,635.3	50.4
2007	341,974	141,723.4	41.4
2008	234,184.7	92,918.09	39.7
2009	361,648.8	141,486.5	39.1
2010	432,561.4	153,977.6	35.6

Source: *Zhongguo guotu ziyuan tongji nianjian* [*China Land and Resources Statistical Yearbook*] (successive years).

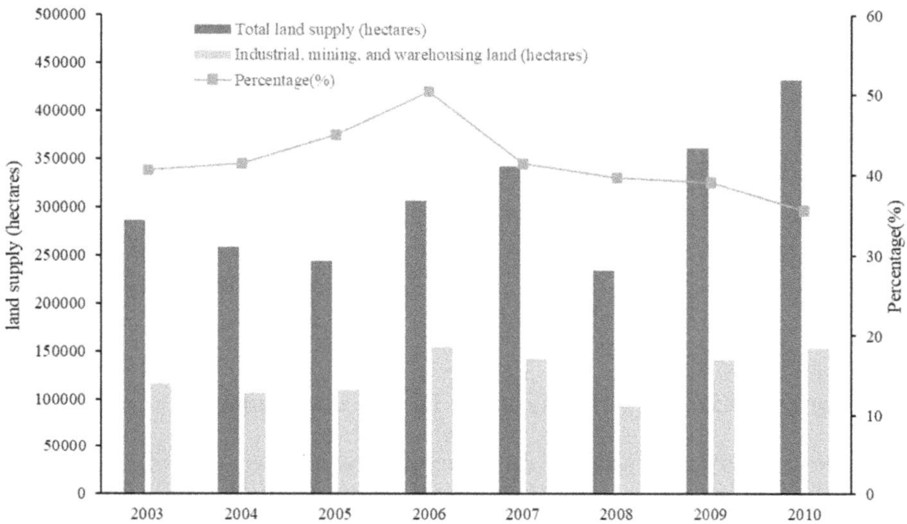

Figure 3.4: Industrial, mining, and warehousing land as a share of total land supply, 2003–2010.

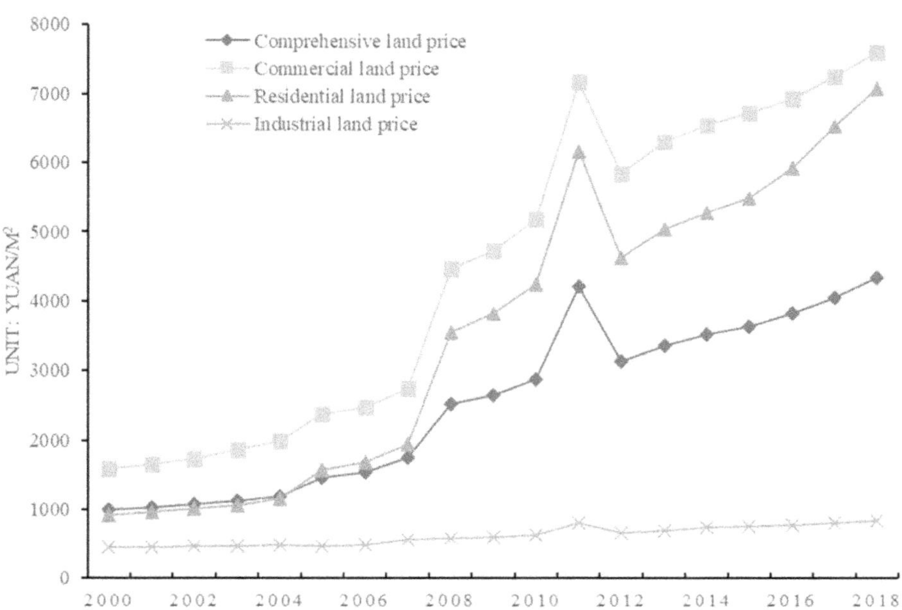

Figure 3.5: Changes in national land price levels, 2000–2011 (yuan/m^2).

Based on the analysis of 21 cities across the country, from 2000 to 2011, industrial land price nationwide increased from 467 yuan/m^2 to 706 yuan/m^2. In the east, industrial land increased from 580 yuan/m^2 to 883 yuan/m^2; in the central region, it increased from 347 yuan/m^2 to 566 yuan/m^2; and in the west, it increased from 391 yuan/m^2 to 560 yuan/m^2. The government's intervention in industrial land prices was conducive neither to the transfer of manufacturing from the east to the central and western regions, nor to the optimization and upgrading of industries in the east and the more intensive use of scarce land (see Table 3.5).

3.3 Rapid Urbanization Driven by Land Expropriation

Since the mid-1990s, China's urbanization process has greatly accelerated. From 1995 to 2010, China's urbanization rate increased at an average annual rate of 1.5%, from 29.04% to 49.95%. This rapid urbanization drove high economic growth, promoted industrial restructuring, and caused dramatic changes in the means of production, lifestyles, and values of rural residents. During this period, each 1 percentage point increase in the urbanization rate brought an increase of 633 yuan in per capita GDP. If we examine the contribution rates of three different kinds of industry to GDP, we see that the contribution rate of primary industry fell from 9.1% to 3.9%, of secondary industry fell from 64.3% to 57.6%, and of tertiary industry rose from 26.6% to 38.5%. In China, a floating population, made up primarily of farmers who go to the cities to work and live, and which accounts for the majority of China's farming population, enjoys the benefits of development and is provided with unprecedented opportunities.

Rapid advancement of urbanization has benefited from China's unique land expropriation system, planning system, and local governments' exclusive land sales.[5] Because the government can expropriate land at low cost and has its exclusive land transfer system, it provides local governments an incentive for expropriating and occupying more farmers' collective land. To this end, the local governments use the planning controls and revision rights and do their most to mark off and bring rural and suburban land into the urban planning circle. They use their legal land expropriation authority to expropriate farmers' collective land and convert it to state-owned land, and then they use their exclusive land sales authority to assign the government-controlled land to the land market, thus maximizing the land's value. As the cities' land area continues to expand, the share of state-owned land increases, and more and more land comes under governmental control.

Table 3.5: Changes in industrial land prices in different regions, 2000–2011.

Region(yuan/m²)	2000	2001	2002	2003	2004	2005	2006	2007	2008	2009	2010	2011
Average industrial land price in 21 cities	467	475	486	495	508	516	536	640	613	634	672	705
Average in east	580	588	598	610	627	639	665	831	769	784	837	883
Average in central region	347	351	364	367	377	382	399	492	508	518	549	566
Average in west	391	400	413	421	430	437	448	471	476	511	535	560

Source: Ministry of Land and Resources, *Tudi shichang jiance dongtai* 《土地市场监测动态》 [*Land Market Monitoring Dynamics*].

In recent years, local governments have rapidly expanded their urban areas through planning revisions and adjustments to administrative divisions. We call this "expansionary urbanization of urban circles" (*chengshijuan waiyi de chengshihua* 城市圈外移的城市化). From 1995 to 2010, the area of built-up urban areas nationwide increased by 20,793.8 square kilometers, when an average annual increase of 1,386.3 square kilometers (see Table 3.6).

From 1999 to 2008, the development of all types of cities has shown rapid expansion from their centers to spaces along their outer edges. During that time, the built-up area for China's top 10 cities expanded from 2,629 square kilometers to 7,727 square kilometers (see Figure 3.6).

In the land urbanization process, land expropriation is the main method, and the forced low price of expropriated land the main incentive. As urbanization progresses, urban maps continue to extend into the countryside, and the collective land of the farmers is expropriated by the government as state-owned. From 2005 to 2010, China's newly added residential, industrial, and mining construction land totaled 21,627 square kilometers, and during the same period, the land expropriated for residential, industrial, and mining construction use totaled 14,551 square kilometers, that is, expropriated land accounted for 67.3% of that newly added land area used for residential, industrial, and mining purposes (see Table 3.7).

Meanwhile, the government uses expropriated land for financing by assigning it and by collateralizing it, in order to raise funds for urban infrastructure construction and provide leverage for rapid urbanization. As the pace of

Table 3.6: China's urban areas and urban built-up areas, 1995–2010 (unit: km^2).

Year	Urban areas	Urban built-up areas	Year	Urban areas	Urban built-up areas
1995	1,171,698	19,264.2	2003	399,173.2	28,308
1996	987,077.9	20,214.2	2004	394,672.5	30,406.2
1997	835,771.8	20,791.3	2005	412,819.1	32,520.7
1998	813,585.7	21,379.6	2006	166,533.5	33,659.8
1999	812,817.6	21,524.5	2007	176,065.5	35,469.7
2000	878,015	22,439.3	2008	178,110.3	36,295.3
2001	607,644.3	24,026.6	2009	175,463.61	38,107.3
2002	467,369.3	25,972.6	2010	178,691.7	40,058.0

Source: *Zhongguo chengshi jianshe tongji nianjian 2008*《中国城市建设统计年鉴 2008》[*China Urban Construction Statistical Yearbook, 2008*].

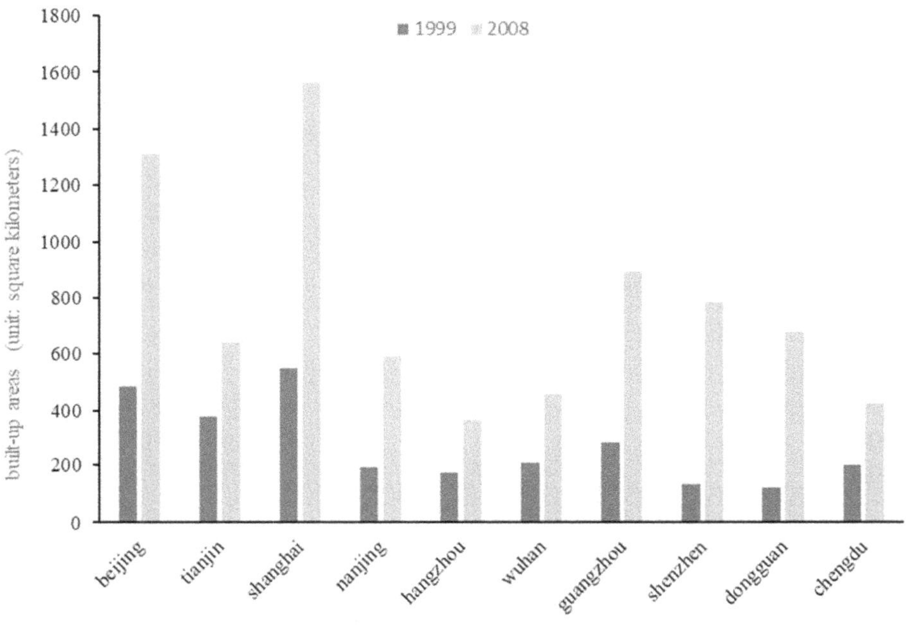

Figure 3.6: Comparison of the built-up areas of major cities in China, 1999 and 2008 (unit: sq. km.).

urbanization accelerates, there is a huge demand for infrastructure. Since 2000, local governments in China have relied on land sales income, land reserve centers, and various financing platforms to obtain large amounts of bank loans with land as collateral, in order to ensure funding for land development and infrastructure building. Funding for urban infrastructure building mainly comes from three sources—government investment, loans, and self-raised funds. In 2008, government investment accounted for 32.2% of urban infrastructure funds, bank loans 29.6%, and self-raised funds from business enterprises 28.7%. Each of these three funding sources are highly dependent on land.

Now, while the proportion of China's urban infrastructure investment from the central government's financial allocation dropped from 26% in the early 1980s to 1.1% by 2008, local governments became the mainstay of government investment. After the tax reform of 1994, which further divided the central and local governments' tax systems (to reflect the shift toward a more market economy), the local governments had already established a pattern of "relying on financing for revenue, and depending on land for construction." In other words, they have become basically dependent on revenue from their land operations to

Table 3.7: The proportion of total newly added residential, industrial, and mining construction land area obtained through land expropriation, 2005–2010.

Year	Area of land expropriated for residential, industrial, and mining construction (10,000 *mu*)	Additional land designated for construction of residential, industrial, and mining	Proportion of additional land designated for construction of residential, industrial, and mining from expropriated land (%)
2005	283.4	486	58.31
2006	329.5	588	56.04
2007	373.4	475	78.61
2008	266	445	59.78
2009	499.5	440	113.52
2010	430.9	810	53.20
总计 Total	2182.7	3244	67.28

Source: *Zhongguo guotu ziyuan tongji nianjian.* 《中国国土资源年鉴》[*China Land and Resources Statistical Yearbook*].

fund all their infrastructure building. As for bank loans and self-raised funding, this comes mainly from local city investment or financing platform companies, but the ultimate source for repayment, again, is mostly tied up with the land. The bank loans are pledged mostly by the land use rights and land revenue rights; they are repaid by future income derived from the land. The Ministry of Housing and Urban-Rural Development estimates that in 2009, the land sales fees used for investment in urban infrastructure building was around 334.1 billion yuan, which accounted for 48% of the total funding for such investment that year (the other sources included direct investment by government and subsidies for such investment under the names of business enterprises). Since the total land sales fees collected in 2009 was calculated to be 1.59 trillion yuan, this means that 21% of the land transfer fees were used for urban infrastructure building. Let us consider the example of urban investment in Chongqing. From 2002 to 2008, the total financing scale was 85 billion yuan, 40% of which came from bank loans, and of this, 63% came from land use rights and income rights allocated by the government as a pledge. In addition, revenue from land consolidation contributed 13% of the total financing.

4. The Economic and Social Consequences of Land Expropriation

4.1 Endangerment of National Food Security Due to Large-Scale Occupation and Use of High-Quality Cultivated Land

Due to the low cost of land expropriation, local governments are engaged in the construction of development zones, cities, and large-scale key projects, and because they usually choose to build on land where development costs are low, large amounts of high-quality cultivated land are expropriated. It has been calculated that from 1996 to 2008, the area of cultivated area occupied by construction totaled 2,703,700 hm^2 (see Fig. 3.7), which resulted in a sharp decline in the area of cultivated land. During the same period, the net reduction in cultivated land area was 8.3171 million hectares, a decrease of 6.40%. The amount of arable land per capita decreased from 0.106 hectares in 1996 to 0.092 hectares in 2008. Moreover, the cultivated land that was occupied was mainly high quality (see Figures 3.7, 3.8, and 3.9).

The rise of industrialization and urbanization along China's coast also came at the expense of the high-quality farmland in these areas. The increased quantity of cultivated land occupied by construction in the eastern and southeastern parts of China after 2000 caused a sharp decrease in the quality of cultivated land, and the center of gravity of cultivated land shifted northeast. Economically developed areas, such as the Jiangsu-Zhejiang-Shanghai area and the Beijing-Tianjin-Hebei area, stand out especially for the quantity of cultivated land now covered by construction. These areas rank first in the nation in terms of total amount of newly added construction land and reduction of construction land. These areas were traditionally China's main food-producing areas, but the occupation of such large quantities of their cultivated land has led to a nationwide change in the food supply-and-demand map. The main food-producing areas moved from the southeast to the north and central western regions, and accordingly, food production shifted from regions with abundant natural rainfall and a warm climate to regions with dry, cold climates. Currently, only six of the 13 major food-producing provinces and autonomous regions in the country are net food exporters. In 2007, China's 13 major food-producing areas accounted for 75% of the country's total food output. These include Hebei, Inner Mongolia, Liaoning, Jilin, Heilongjiang, Shandong, and Henan, which are in the north. Their share of the total national food production was 36.2% in 1991, but by 2007 it had increased to 43.5%. Meanwhile, total food production in the south

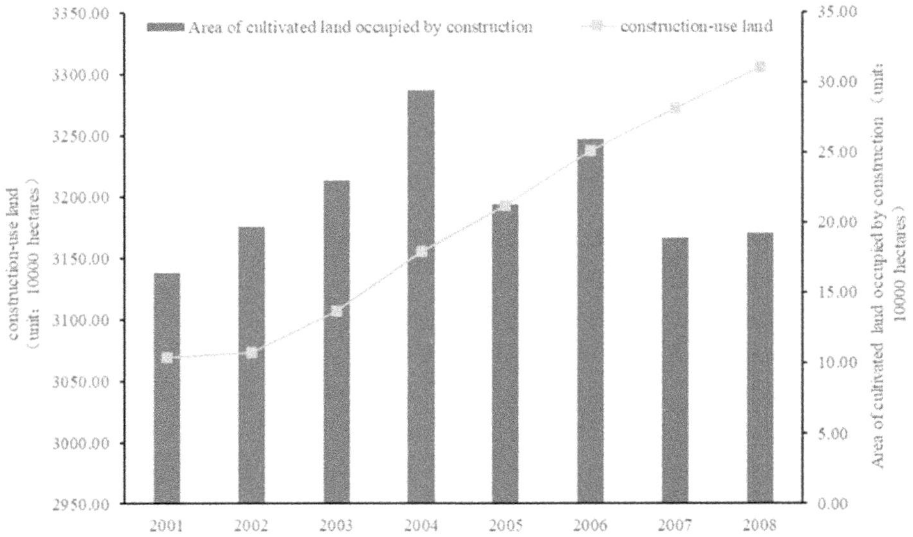

Figure 3.7: Schematic diagram of total national construction land and occupied cultivated land, 1996–2008.
Source: Ministry of Land and Resources Land Change Survey data.

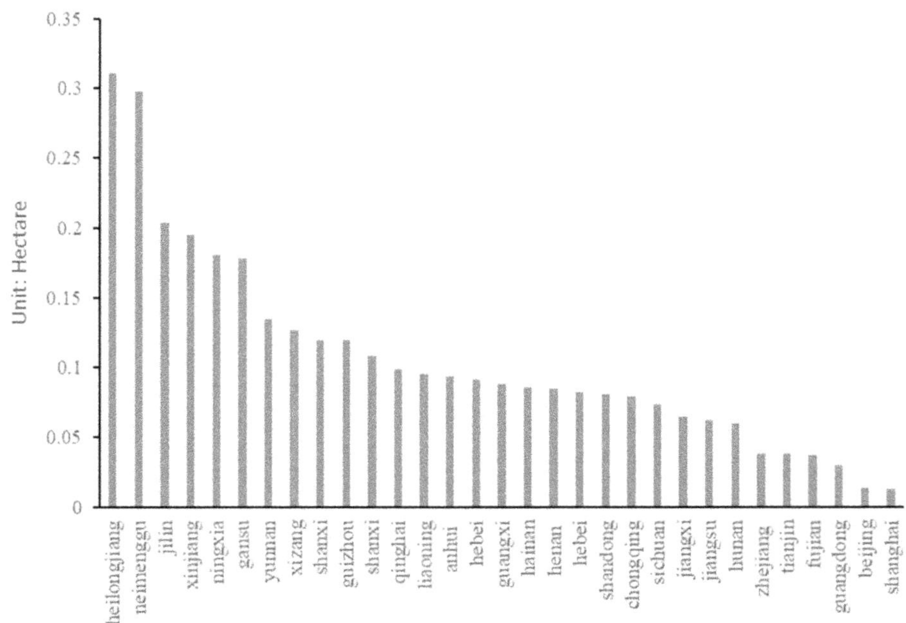

Figure 3.8: Per capita cultivated land in each province.

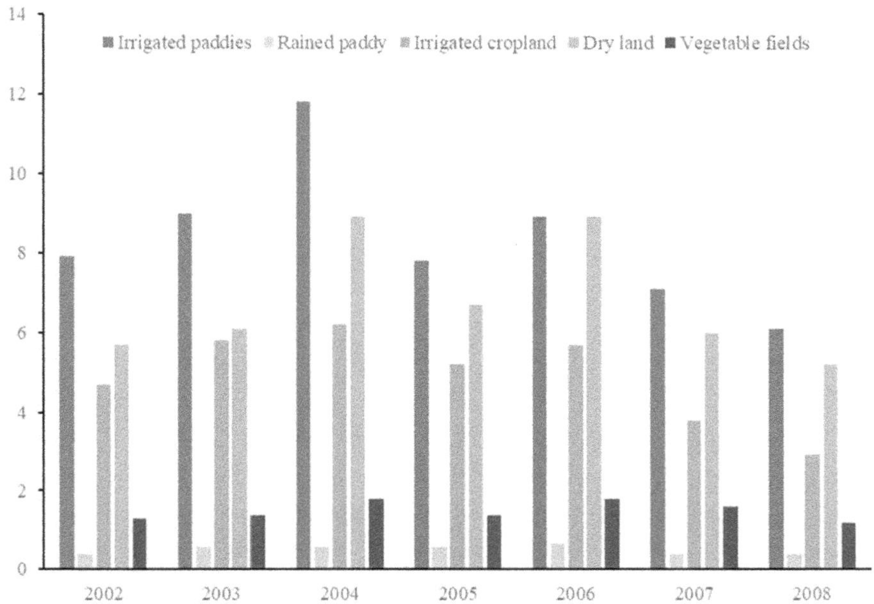

Figure 3.9: Area of cultivated land, by category, occupied by construction, 2002–2008.

declined. The six main food-producing regions in the south—Jiangsu, Anhui, Jiangxi, Hubei, Hunan, and Sichuan—saw their share of total national food production fall from 36% in 1991 to 31.6% in 2007. In the seven main consumption areas—Beijing, Tianjin, Shanghai, Zhejiang, Fujian, Guangdong, and Hainan—the food production-demand gap has expanded each year. Their share of total national food production fell from 12.2% in 1991 to 6.3% in 2007, so that the production-demand gap in these regions increased from 48.5 billion kg to approximately 55 billion kg in 2007. We should be even more vigilant of the fact that, with the accelerating industrialization of the central and western regions, land use and expropriation are still replicating the development pattern of the eastern region—and the pace of land expropriation is accelerating. The threat to China's national food security cannot be underestimated.

4.2 Reduced Governmental Credibility Due to Overstepping Its Role

The conversion of land from agriculture is completed mainly by the government, but in the land expropriation process, it is the local government that plays the role of decision-maker, fixer, organizer, implementer, and arbiter. Since building cities and development zones depends primarily on land sales and land mortgage financing, a local government also has an incentive to more land expropriation, more land tenure, and more land sales. This has resulted in a large number of occurrences of illegal land use by local governments. In 2010, there were 66,000 illegal land use cases nationwide, involving a total of 677,000 *mu* and the occupation of 270,000 *mu* of cultivated land. Cases involving the construction of railways, highways, water conservancy, and other infrastructure projects without prior approval were prominent among them. In 2009, illegal land use by national and provincial key projects accounted, respectively, for 45% and 55% of the land area involved in that year's cases of illegal land use and illegal occupation of cultivated land. The problem of local governments leading, supporting, or acquiescing in the illegal supply of land still occurs from time to time.

Judging from the 127 instances of land expropriation disclosed by the media since 2003, most of the land-expropriated farmers' appeals were prompted by the expropriated land's use for commercial purposes. Of the 127 land expropriations, 31 were used for commercial building construction, 33 for industrial park construction, and 10 for commercial tourism development; these three situations accounted for 58% of the reported incidents. The reason why land-expropriated farmers reacted so strongly to the land's conversion to commercial use was the low compensation they received from the government contrasted too greatly with the high fees it received for assigning it as commercial land. Meanwhile, mass incidents caused by land expropriation for key projects and infrastructure-building accounted for a rather large proportion—38.6% of the reported incidents. These included 20 road constructions, 7 railway constructions, 3 airport constructions, 8 parkland and greening projects, and 11 science, education, culture, and health facility projects. Of the 127 instances reported in the media, most of the land expropriations that triggered appeals by farmers and mass incidents involved some unregulated or illegal land expropriation behavior by local governments. There were 65 cases, accounting for 51.2%, in which the local governments did not provide compensation according to law, and 53 cases, accounting for 41.7%, in which the land was occupied and used without prior approval.

Field surveys found that in order to expropriate land and demolish property, some local government officials had to "rush to the front" and take the farmers' land, uproot their crops, tear down their houses, dig up their ancestors' tombs, cut their trees, and generally "charge in and capture their positions" (*chongfeng xianzhen* 冲锋陷阵). This not only seriously damaged the image of government officials in the minds of farmers, but also caused a great reduction in the government's credibility. To defend their land, farmers' collectives had to put up resistance, and thus a confrontation between farmers and the government was formed. At this point, it became relatively common for the local governments to mobilize the police forces in order to proceed with the land expropriation. This abuse of the police forces directly intensified the social conflicts and became the fuse for mass incidents. From the details of these events disclosed by the media, the land-expropriated farmers' dissatisfaction was due to a combination of many factors. The causes of conflict in the 127 incidents were due mainly to the fact that the compensation for land expropriation was too low; this was the case in 107 of the incidents, putting it in first place. Other factors were lack of clarity in the division of payment for land (34 incidents), farmers' not knowing that their land would be expropriated (32), farmers not being told what the land would be used for (23), and misappropriation of compensation (20). However, once a local government responds inappropriately to farmers' dissatisfaction and appeals, it exacerbates the conflict and even causes it to evolve into a mass incident. In 81 of the 127 land expropriations, farmers reported that the cause of conflict was local governments and construction units' violent behavior—which makes this the second leading cause of the reported incidents (see Figure 3.10).

As urbanization progresses, the value of land becomes more and more obvious. The value-added income from land climbs rapidly, so the competition of land interests steadily intensifies. Hence land-related disputes have become an important reason why people lodge appeals to the government. At present, the scope of distribution of land interests in China has expanded and deepened somewhat, so the problems have become more and more prominent, but the most important one has been the conflict of interests between land-expropriated farmers and local governments. In the current distribution of land benefits, the land-expropriated farmers do receive land compensation and resettlement subsidy, and in recent years the state has also established a social security system for land-expropriated farmers. However, in general, the land-expropriated farmers have not had a share in the results of urbanization. Furthermore, violations of the land expropriation procedure, deductions from land acquisition payments, and too-low compensation standards for land expropriated for state key projects have caused dissatisfaction among farmers in some places, and this makes land expropriation a prominent problem affecting China's social stability. According to statistics of the National Public Complaints and Proposals Administration, 60% of the collective appeals they receive are related to land. Land disputes have become the number one focus of farmers' appeals since tax reform and account for 40% of all appeals received, and of these, 84.7% relate to disputes over compensation for land expropriation. Each year, approximately 4 million disputes are caused by

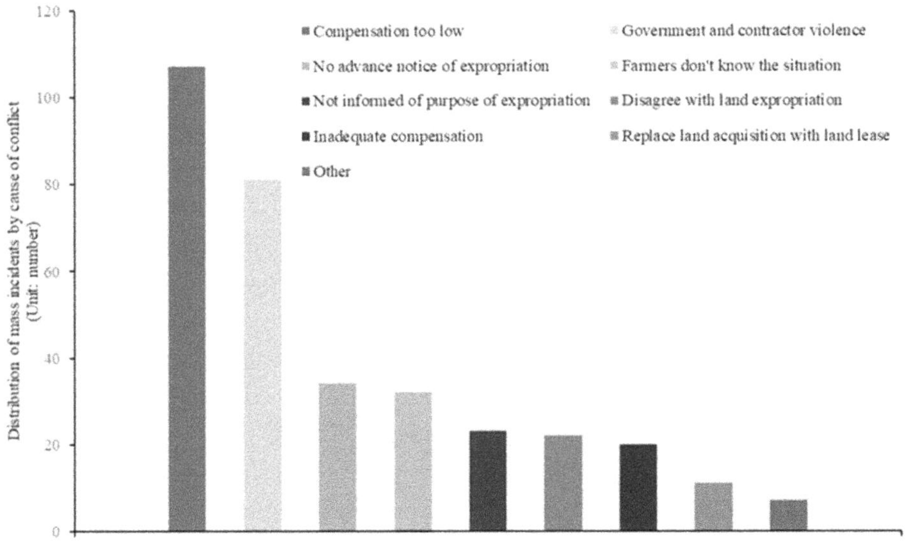

Figure 3.10: Distribution of mass incidents by cause of conflict.

land expropriation and demolition. Expropriation-driven development inevitably leads to conflicts between farmers and the government, until opposition between them forms.

Among the 127 media-reported land expropriation cases, the government improperly handled the appeals and complaints lodged by farmers, which intensified the conflicts, formed opposition between the government and farmers, and became the main cause of mass incidents. The government resorted to extreme methods in 95 of the incidents. Among these, 69 involved mobilization of police, armed police, and other exclusive governmental tools; 48 involved arrests; 45 involved beatings; 23 involved gang inventions; and 3 even involved gunfire. In only 36 incidents were compromises made to increase the compensation, and in another 6 cases, a divide-and-conquer approach was used. In 27 cases, the government simply ignored all appeals. Only 15 of the cases were decided in court (see Figure 3.11).

4.3 Debt Risk for Local Governments and Financial Risk for the Central Bank

The land expropriation-for-development model has ensured the rapid growth of China's economy, but it has also led to the Chinese economy's excessive

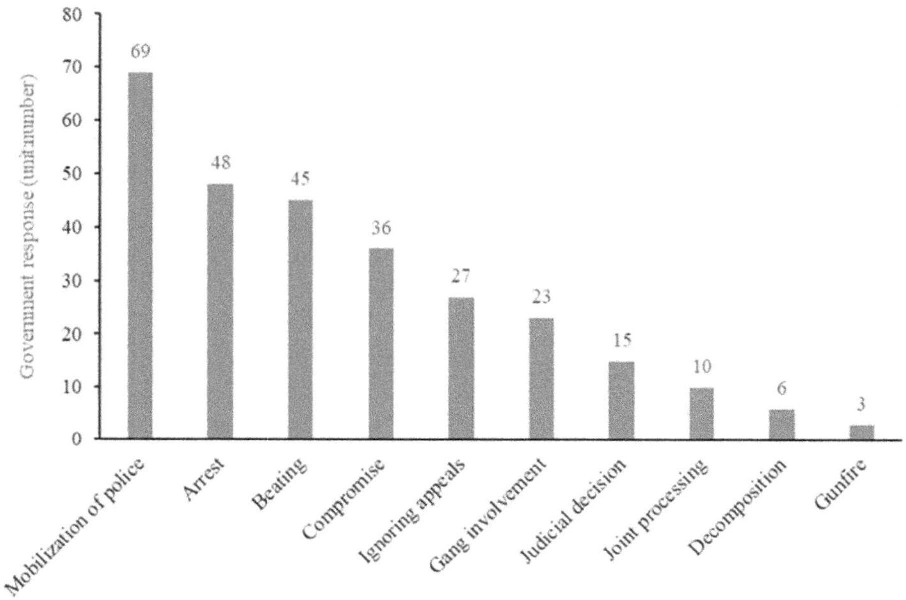

Figure 3.11: Government response in land expropriation conflicts.

dependence on land, resulting in local government debt risks and bank financial risks from land sales and the use of land financing. Since 1997, local government debt in China has increased year by year. In 1998 and 2009, their debt balance had increased, respectively, by 48.20% and 61.92% over each of the previous years. In 2010, the debt balance increased by 18.86% over the previous year (see Fig. 3.12). There are 78 municipal and 99 county-level governments that have debt-repayment obligation ratios above 100%; these governments account for 19.9% and 3.56% of their respective tiers of government. The debt balance announced by the National Audit Office for 2010 amounted to 10.7 trillion yuan, accounting for 26.9% of GDP for that year. If we add to that public sector debt shown by nominal indicators (about 20% of GDP) and financial bonds issued by policy financial institutions (about 6% of GDP), China's overall public sector debt ratio should be between 50% and 55% (see Figure 3.12).

The local debt situation in some areas is even less optimistic, and problems of imbalance and unsustainability are quite prominent. At the end of 2010, the debt balances of the 11 provinces (municipalities) and five municipalities with independent planning status (*jihua danlie shi* 计划单列市) in the eastern region amounted to 5.320839 trillion yuan, accounting for 49.65% of the total; the debt

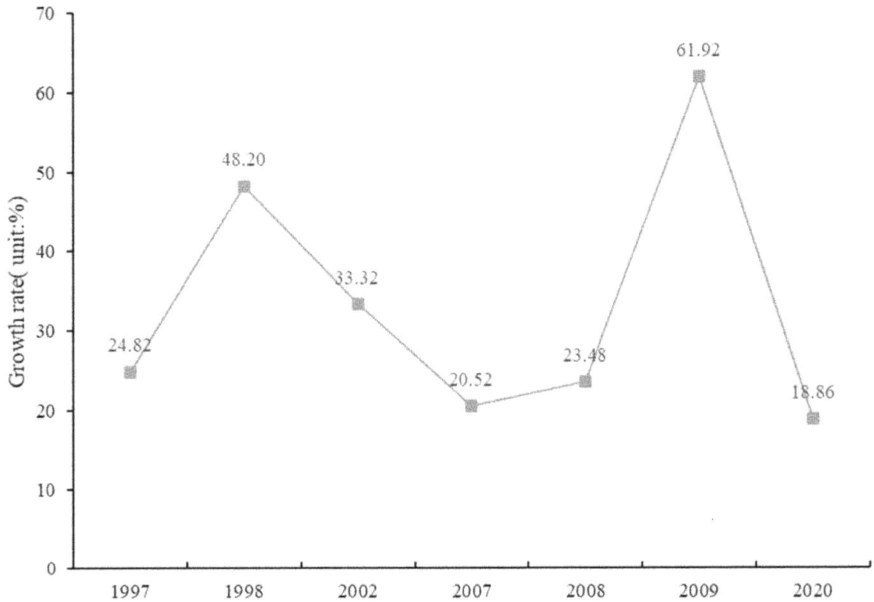

Figure 3.12: Growth rate of local government debt balance.
Source: National Audit Office, *Quanguo defang zhengfuxing zhaiwu shenji jieguo* 《全国地方政府性债务审计结果》 [*Results of National Audit of Local Government Debt*], June 2011.

balances of the eight provincial governments in the central region amounted to 2.471635 trillion yuan, accounting for 23.06%; and the debt balances of the 12 provinces (autonomous regions, municipalities) in the western region amounted to 2.925017 trillion yuan, accounting for 27.29%. At the end of 2010, the municipalities with independent planning status in the east and the central region and western region had debt ratios of 19.2%, 24.2%, and 36.4%, respectively. What should draw our attention, however, is that collateralized land has become the most important source of revenue for governments to repay their debts, and that the repayment of local debts depends more heavily on income from land sales. An investigation by the National Audit Office showed that at the end of 2010, except for 54 county-level governments, all provincial- and municipal-level governments and the vast majority of county-level governments had government liabilities, and that the amount of land sales income they had promised for the repayment of debts amounted to 2.547351 trillion yuan, which accounts for 37.96% of the amount of debt they are obligated to pay off. If one day the income from land sales in various places were to decline, the local governments would quickly become insolvent.

To build infrastructure, local governments have set up all kinds of financing platforms and make use of the state-owned land in their hands as collateral for huge bank loans. By now, these local financing platforms have formed a cycle: collateralize land → obtain loans → acquire more land → collateralize land. According to the National Audit Office's findings, as of the end of 2010, the debt balance of these financing platform companies amounted to 4.971068 trillion yuan, accounting for 46.38% of total local government debt balance. Of the debt balance already spent by the local governments of various levels, 1.020883 trillion yuan, or 10.62%, was used for land acquisition and reservation.

If we examine the sources of local government debt, we find that banks are the largest providers of funds. As of the end of 2010, banks had loaned local governments 8.467999 trillion yuan, accounting for 79.01% of local government debt funding. Of this amount, the debt which the governments bear the responsibility to repay amounted to 5.0225 trillion yuan, accounting for 74.84% of the total. The proportion of local financing platform loans as a share of total bank loans is unevenly distributed among different types of institutions (see Fig. 3.13). At the end of 2009, loans to local government financing platforms by major listed banks accounted for about 10% of the total loans, the proportions of loans from China Development Bank (CDB) and local financial institutions were higher. Among the latter, loans from CDB to the financing platforms (with capital construction loans as the statistical standard) accounted for as much as 69% (see Figure 3.13).

The local financing platforms repay a relatively high proportion of their debts via the method of "taking out new loans to pay off the old." At the end of 2010, 358 financing platform companies used precisely that method to repay government debts and other related debts of 105.971 billion yuan, so that 55.20% of the repayments, on average, were made this way. Included here are 387 colleges and universities and 230 hospitals, where more than 50% of the repayment on debt for which the government bears responsibility, and on other related debts, was made by "taking out new loans to pay off the old." In 2011, the China Banking Regulatory Commission (CBRC) explicitly prohibited this method, which made the short-term debt repayment risk of the financing platforms even more conspicuous.

The overdue debt ratio of local financing platforms will further increase. At the end of 2010, 148 financing platform companies, or 2.3% of the total, had overdue debts of 8.004 billion yuan, and the overdue debt ratio averaged 16.26%. If we consider that in the next few years the local governments' income from land sales and mortgage prices is expected to decline, we can see that this "taking out

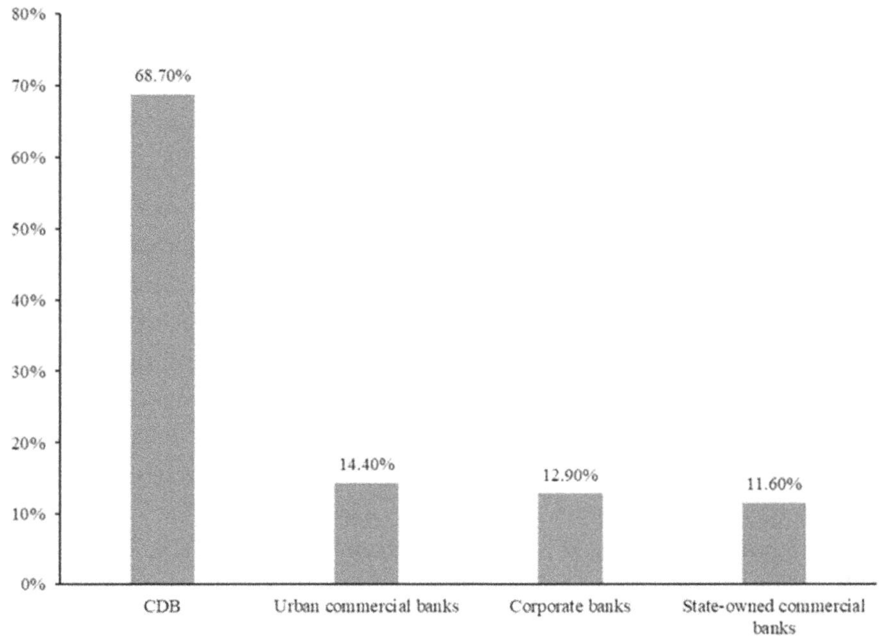

Figure 3.13: Balance of loans from various financial institutions to financing platforms as a percentage of total loans.
Note: Statistical standard for loans from CDB.
Source: Bank annual financial reports.

new loans to pay off the old" method is inherently unsustainable. Hence in the next few years the overdue debt ratio is expected to increase.

The local financing platforms have rather long debt cycles and relatively weak profitability. If we examine the yearly debt repayments, we see that, with regard to the debt balances of local government at the end of 2010, 24.49% and 17.17% of the debt was repaid in 2011 and 2012, respectively, and in 2013, 2014, and 2015, 11.37%, 9.28%, and 7.48%, respectively; repayments in 2016 and after accounted for 30.21%. Since the funds borrowed by the financing platform companies are invested mainly in public interest or quasi-public interest projects with long payback periods and relatively poor ability to have sustained cash flow, in 2010, for example, 26.37% of the total 1,734 financing platform companies suffered losses (see Figure 3.14).

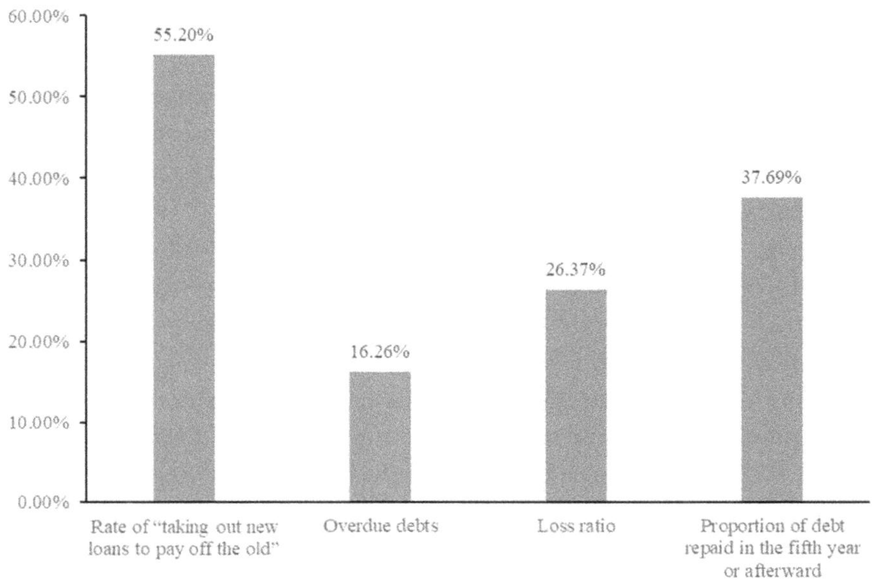

Figure 3.14: Bank risks on loans to local financing platforms.
Source: National Audit Office, *Quanguo difang zhengfuxing zhaiwu shenji jieguo* 《全国地方政府性债务审计结果》 [*Results of the Audit of Local Governmental Debts Nationwide*], June 2011.

4.4 Unfair Distribution of Land Value-Added Income

Local governments have become the main recipients of land value-added income because they can legally force the expropriation of farmers' land at a low price and then have the exclusive right to assign use of that land. From 1999 to 2009, the ratio of the transaction value of revenue from their land sales to total revenue rose from 0.092:1 to 0.437:1, and the ratio of the transaction value of revenue from their land sales of state-owned land to their extra-budgetary revenue rose from 0.163:1 to 1.942:1. The role of land sales income in local development is becoming more and more significant. Since the 2008 financial crisis, local governments' dependence on land financing has not decreased at all. Instead it has intensified. City governments in particular have grown even more dependent on land financing (see Table 3.8).

Because of the dual landownership systems—state ownership of urban land and collective ownership of rural land—and the fact that any unit needing land must use state-owned land, most of the land value-added revenue goes to city governments, and most of the land sales income is used by the cities, too. This results in an urban-rural imbalance in the use of land sales revenue. At present, the part used by the cities is used mostly for urban construction, the State Land Revenue Fund (国有土地收益基金), and affordable housing projects. From 2008 to 2010, an average of 58.76% of land sales revenue went to urban construction (66.27%, 48.64%, and 61.65% for the three years, respectively), an average of 7.43% of the revenue went to the State Land Revenue Fund (6.77%, 6.41%, and 8.25%, respectively), and an average of 3.37% of the revenue went to affordable housing projects (3.19%, 2.72%, and 3.80%, respectively). Total expenditures for these three items for the three years—76.23%, 57.78%, and 73.70%—were an average of 69.56% of the land sales revenue.

Meanwhile, the percentage of land sales revenue that goes to rural villages and agriculture is still low. Most of it goes to the Agricultural Land Development Fund (农业土地开发资金) and the Rural Infrastructure Building Fund (农村基础设施建设资金). From 2008 to 2010, an average of 1.94% of the revenue from land sales went to the Agricultural Land Development Fund (2.76%, 2.08%, and 1.55% for the three years, respectively), and an average of 7.58% of the revenue went to the Rural Infrastructure Building Fund (7.41%, 6.31%, and 8.36%,

Table 3.8: Ratio of transaction value of income from sales of state-owned land to local fiscal revenue and extra-budgetary revenue, 1999–2009 (billions of yuan).

Year	Transaction value	Local government fiscal revenue	Local government extra-budgetary revenue	Ratio of land sales income to local government fiscal revenue	Ratio of land sales income to local government extra-budgetary revenue
1999	51.43	559.487	315.472	0.092	0.163
2000	59.56	640.606	357.879	0.093	0.166
2001	129.59	780.33	395.3	0.166	0.328
2002	241.68	851.5	403.9	0.284	0.598
2003	542.13	984.998	418.743	0.55	1.295
2004	641.22	1189.337	434.849	0.539	1.475
2005	588.38	1510.076	514.158	0.39	1.144
2006	807.76	1830.358	594.077	0.441	1.360
2007	1221.67	2357.262	628.995	0.518	1.942
2008	994.39	2770.342		0.359	
2009	1423.97	3258.074		0.437	

Source: *Zhongguo guotu ziyuan tongji nianjian*《中国国土资源统计年鉴》*[China land and resources statistical yearbook]* (successive years).

respectively). Total expenditures in rural areas for these two items for the three years—10.18%, 8.39%, and 9.91%—were an average of 9.52% of the land sales revenue.

Following the implementation of policies to increase the support of land sales income to agricultural and rural areas, accrue education funds, farmland water conservancy construction funds, and so on, the structure of land sales revenue and its expenditure will change accordingly, and the proportion of land sales revenue spent on urban construction is expected to gradually decrease. However, the share of land sales income that gets allotted to farmers is unreasonable. In the past five years, even though the standard rates of compensation for land expropriation in various places have increased by 30% or more, their share of the land sales income has not—in fact, it has decreased substantially. According to statistics on compensation for land expropriation and land sales income for Liaoning, Zhejiang, and Henan provinces, in 2007, compensation fees for land expropriation in the three provinces accounted for 16.82%, 8.62%, and 19.58% of their land sales income, but in 2010, those percentages had dropped to 10.37%, 2.91%, and 12.99%, or by more than one-third.

4.5 Increased Social Risk

We analyzed the 127 land expropriation appeals and mass incidents publicly reported by the domestic media since 2003, and the seriousness, likelihood, and quantity of the expropriation-triggered mass incidents show an upward trend. Chronologically, since 2003, the frequency of conflicts and mass incidents touched off by land expropriation has increased. Among the 127 reported incidents, there was a significant increase after 2007, with 45 incidents occurring in 2010. Mass incidents from land expropriation occurring in 2010 and 2011 accounted for 62% of the reported incidents in the 2003–2011 period (see Figure 3.15). Regionally, land expropriation conflicts were distributed broadly, but the more economically developed the area, the more concentrated the conflicts. The 127 land expropriation conflicts were distributed in 27 provinces (municipalities, autonomous regions) and 113 counties (districts). Of these, 24 conflicts occurred in 17 counties (districts) in Guangdong, the most economically developed province. The next leading province was Henan, with 12 conflicts in 12 counties (districts). After this was Hebei, with 9 conflicts in 9 districts; Beijing, with 9 conflicts in 8 districts; Jiangsu, with 8 conflicts in 7 counties and districts; and Zhejiang, with 7 conflicts in 4 countries and districts (see Figure 3.16).

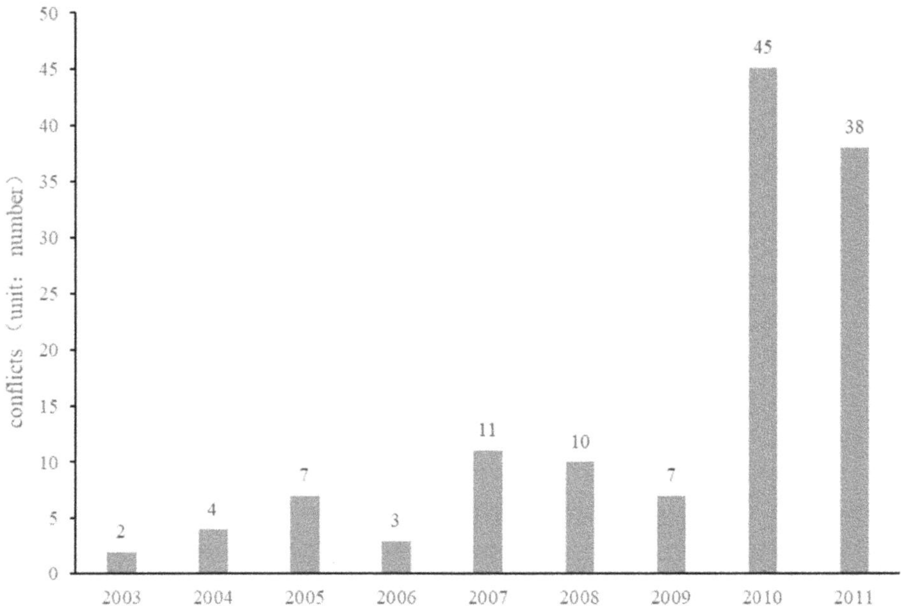

Figure 3.15: Distribution by year of media-reported land expropriation conflicts since 2003.

When faced with land expropriation, farmers usually make collective appeals. Once the local government cannot properly resolve the issue, the appeal moves up to the next level. From the incomplete statistics based on media reports, appeals were made in 52 of the cases, and of these, 43 were collective appeals. Farmers resort to collective appeals mainly because land expropriations generally involve the land of one village or even several villages, which affects the livelihoods of the ordinary people living in those areas. To resolve the land expropriation problem, the farmers will first make appeal to the town or village government. Among the 43 collective appeals, 11 were first presented to the town or village, and of these, 9 were moved up to the district level, of which 8 were moved up to the city level, and of these, 3 were even moved up to the provincial level, and 2 up to the national level. As for the scope affected by the incidents, 83 affected the village and township level, 77 affected the district level, 59 the city level, 21 the provincial level, and 8 the national level. When interest claims are not satisfactorily met, the land-expropriated farmers tend to resort to extreme methods. These methods are increasingly jeopardizing social stability, and the resulting social repercussions are growing.

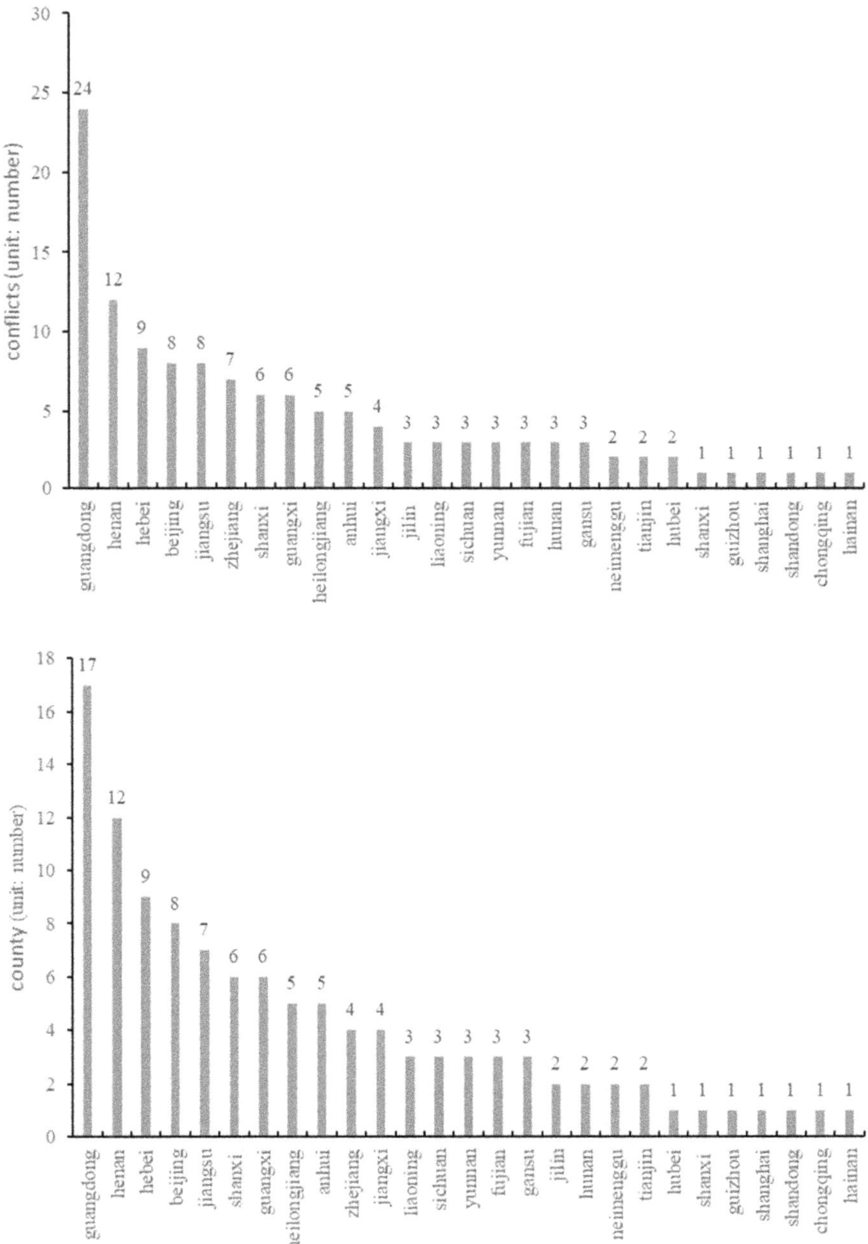

Figure 3.16: Geographic distribution of land expropriation conflicts since 2003 (by province and county).

Among the 127 reported land expropriation conflicts, there were 72 in which farmers resorted to extreme methods. In 39 of them, they laid siege to government buildings or resorted to armed confrontation; in 18, there were sit-in demonstrations; in 14, they blocked roadways; in 43, they stopped construction or vandalized the construction project; in 17, they refused to cooperate; and in 11, they resorted to the most extreme and helpless method of suicide by self-immolation. Very few chose to take proper action: only two cases chose tip-offs (ju bao 举报); five were reported to the police; and 12 were brought before the court. The details of conflict incidents reveal that the reason farmers do not choose the proper ways is mainly because those approaches can scarcely resolve farmers' interest appeals. Therefore, they adopt improper means to expand their influence and put pressure on local governments to get the solution they want. However, the improper approaches generally affect large numbers of people, last a long period of time, and involve extreme measures that easily heighten the conflict and create uncontrollable situations, which have a great impact on social stability. From the data we have been able to collect, the number of people involved in these conflicts is at most 20,000. There were 58 cases with specific records of casualties, with a total of 25 cases of death. The largest number of casualties in a conflict was 157, and the largest number of deaths in a conflict was 6 (see Figure 3.17 and Table 3.9).

Judging from the experiences and lessons from the previous round of development, the current land expropriation model has led to farmers losing their land as well as their right to develop it, so that they have become the losers in the use of land to plan development. The unfair distribution of land value-added revenue has become the focus of social conflicts. The government's forced expropriation of land and forced demolition of farmers' housing have caused tensions and even opposition between the government and the farmers. This sort of government-led land expropriation and demolition has also caused urbanization costs to climb steadily, as manifested in the steadily rising costs to cities of demolishing illegally constructed buildings. Farmers' resistance to land expropriation delays projects as they are implemented, resulting in the loss of development opportunities. As the number of landless farmers increases, the frequency of land expropriation conflicts also increases, as well as the cost of maintaining stability.

As industrialization and urbanization have accelerated, the scale of land expropriation has expanded rapidly, resulting in an increasing number of farmers who have completely lost their land around the towns and suburbs. As the number of farmers without land, work, and social protections has gradually increased, the hidden dangers of social unrest have also increased. Meanwhile, the farmers' awareness of rights and of safeguarding rights has improved, so that wherever

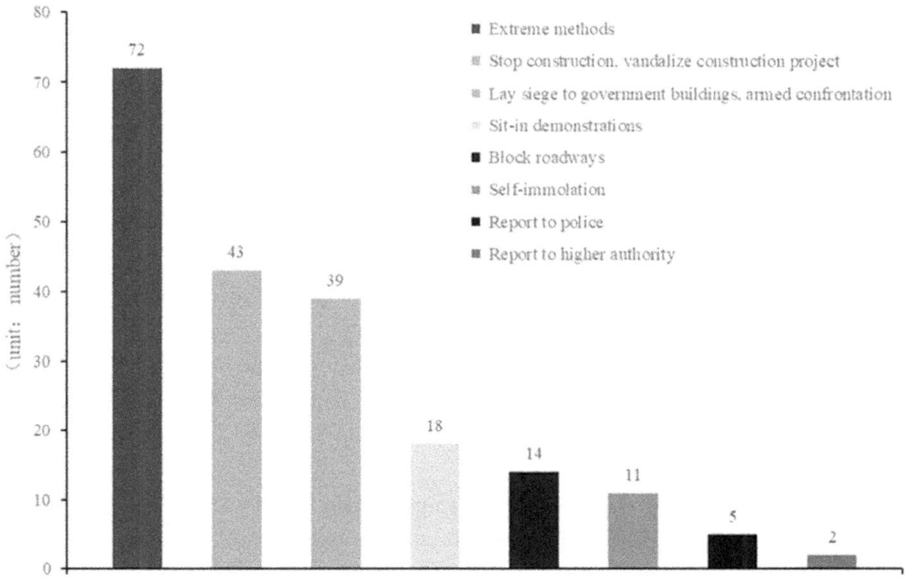

Figure 3.17: Farmers' responses to land expropriation.

Table 3.9: Severity of land expropriation conflict incidents.

Number of incidents with casualty counts	58
Number of incidents with fatality counts	25
Greatest number of injured persons in an incident	157
Greatest number of fatalities in an incident	6

the government expropriates land, that is where they put up a struggle, and their forms of expression and struggle are becoming more and more extreme, affecting more and more people, lasting longer and longer, and directly endangering social stability. As society transforms, social harmony has become an issue of importance equal to that of development, and the cost to the government of maintaining stability is getting higher and higher.

China's current land expropriation system plays an important supporting role in guaranteeing national economic construction and accelerating industrialization and urbanization. However, with the sharp increase in land value-added revenue and the increasingly unfair distribution of land benefits, the low-cost, compulsory land expropriation model has caused social instability, which affects social harmony and endangers the country's long-term stability, and it has reached an unsustainable level.

Notes

1 Liu Shouying刘守英, Zhou Feizhou 周飞舟, and Shao Ting 邵挺, *Tudi zhidu gaige yu zhuanbian fazhan fangshi* 《土地制度改革与转变发展方式》 [*Land System Reform and Methods of Transitional Development*] (Beijing: China Development Press, 2012,P1–10).

2 See also Chen Xizhen 陳錫禎, *Taiwan zhi tudi zhengshou* 《台灣之土地徵收》 [*Land Expropriation in Taiwan*], Taiwan dizheng jianjie congshu 台灣地政簡介叢書 [Introduction to Land Administration in Taiwan Series] (Taipei: Taiwan dizheng yanjiusuo, 2010).

3 Huang Xiaohu 黄小虎, "Zhengdi zhidu gaige de lishi huigu yu sikao 征地制度改革的历史回顾与思考 [Historical Reflections and Thoughts on the Reform of the Land Expropriation System]," *Shanghai guotu ziyuan* 《上海国土资源》 [*Shanghai Land and Resources*] 32.2 (2011.2): 7–13.

4 When there are no special circumstances, the land sales in this chapter refers to the sale of land-use rights.

5 *Translator's note:* It is important to remember that when "assignment" (*churang* 出让) is used in conjunction with land, it refers to the assignment of the right to use land.

4

China's Land Sales Income and Expenditure and Land Finance

In recent years, income from land sales has grown swiftly. At the same time, the central government and relevant departments have continuously introduced policies to strengthen the management of land sales expenditures and adjust the ways in which land sales revenue is spent. Concerns about how the land sales income is spent have steadily increased throughout society and generated widespread debate. This chapter will make a comprehensive analysis of the changing growth trends in land sales income and their causes, the costs and revenue of land sales, the expenditure structure of land sales revenue, policy on land sales income, expenditure and distribution, and problems in managing land sales income and expenditure. The content of "land finance," which is being heatedly discussed in society, is clearly defined, and on this basis, policy recommendation for improving the management of land sales income and expenditure will be put forward.

1. Characteristics and Causes of Land Sales Income Growth

Land sales income is the total land price obtained by the government for allocating the right to use state-owned land by way of sale. This includes: (1) land sales income obtained by bid tendering, auction, quotation, and agreement, and supplemental income obtained for changing the original land use; (2) land lease income collected from lessees of state-owned land that has been leased out according to law, as well as land income that should be paid for the houses on all leased and allocated land; and (3) compensatory for land expropriation and demolition paid to municipal and county people's governments in accordance to law by land users who obtain the right to use state-owned land by way of allocation.

In the land sales income that has been published, two kinds of numeric measures have been used. The first is the total measure of land sales income, that is, numerical data after all the forms of land sales income listed above are taken into account, and the other is the land sales prices, which in comparison with the first, subtracts income from the rental of state-owned land and the fees collected for land expropriation upon allocation (i.e., the second and third income streams listed in the previous paragraph). Since the first kind of numeric measure has only been published since 2008, and the second is the land sales income, which only includes the land sales prices, this chapter will use the income of land sales for land sales prices when analyzing the characteristics of land sales income growth and change, and when analyzing land sales income, expenditure, and costs, the total measure will be used.

1.1 Characteristics of Land Sales Income Growth

1. Income from land sales has changed in stages. With the deepening of social and economic development and the reform of the land use system, the supply of land has increased substantially. Moreover, land sales income shows a phased characteristic, since three significant increases in land sales income have occurred.

The first phase (1989–2000): The *Constitution*, amended in April 1988, and the subsequently revised *Law on Land Administration* added the article, "The right to use land may be transferred in accordance with the law." Thus the transfer and resales of land use rights were launched nationwide. Income from land sales increased from 447 million yuan in 1989 to 50 billion yuan in 1992, an increase of more than 100 times in only three years. From 1993 to 2000, however, it stabilized at around 50 billion yuan. During this period, the supply of land was mainly by allocation. For example, in 1992, land allocated for use accounted for 97.2% of the total land supply. However, paid land supply was mainly based on contracts, and even in Shenzhen, which pioneered paid land sales, the income from land auctions in 1998 accounted for only 3% of its income.

The second phase (2001–2006): In 2001, the "Circular of the State Council on Strengthening the Management of State-owned Land Assets" (《国务院关于加强国有土地资产管理的通知》; no. 15) proposed to "vigorously carry out bid invitation and auction," and in 2002, the Ministry of Land and Resources issued its No. 11 directive, "Rules of bid invitation, auction or listing for sales of State-owned Land Use Rights" (《招标拍卖挂牌出让国有土地使用权规定》), which clearly specified four types of for-profit land use (commercial, tourism, entertainment, and real estate) for which the bid invitation, auction or listing (*zhao pai gua* 招拍挂) method would be adopted. Land sales income then reached a new level, jumping from 59.5 billion yuan in 2000 to 542.1 billion yuan in 2003. From 2004 to 2006, it fluctuated around 600 billion yuan. During this period, the proportion of paid land supply increased steadily, and the proportion of land assigned through bid tendering and auction increased rapidly. For example, in 2004, the proportion of paid land sales had already reached 70%, and 29% of that land area had been assigned through bid tendering, auction, and quotation, while the income from bid tendering, auction, and quotation accounted for 56% of the total income.

The third phase (2006 to the present): In 2006, the "Circular of the State Council on Issues Relating to the Strengthening of Land Control" (《国务院关于加强土地调控有关问题的通知》; No. 31) required that "Industrial land must be assigned by means of bid invitation, auction, and listing, and the price of land sales shall not be lower than the published minimum price standard." The *Real Right Law* promulgated in March 2007 stipulated: "In the case of profit-oriented land having industrial, commercial, tourism, entertainment, and business residence use and in the case of a single land parcel having two or more entities that intend to use it, the land shall be assigned by means of public competition on price, such as bid invitation and auction." The comprehensive deepening of the bid invitation, auction, and listing system has significantly increased the marketized allocation of land resources. Land sales income more than tripled, from 807.7 billion yuan in 2006 to 2.71 trillion yuan in 2010. During this period, the proportion of paid land use stabilized at around 70%, and the proportion handled by bid tendering, auction, and quotation further increased, so that in 2010, the proportion of paid land supply reached 68.2%, and of that, the bid invitation, auction, and listing area accounted for 88.3% of the total area of land assigned nationwide. Bid invitation, auction, and listing income accounted for 96% of total land sales income.

2. Land sales income has grown swiftly in recent years. From 2001 to 2010, income nationwide from the sales of state-owned land increased from 129.6

Table 4.1: Income from sales of state-owned land nationwide, 2001–2010.

Year	State-owned land sales income (billions of yuan)
2001	129.589
2002	241.679
2003	542.131
2004	641.218
2005	588.381
2006	807.764
2007	1,221.672
2008	1,025.98
2009	1,717.953
2010	2,819.77

Source: *ZhongGuo Guotu ziyuan tongji nianjian《中国国土资源统计年鉴》[China Land and Resources Statistical Yearbook].*

billion yuan to 2.8198 trillion yuan, an increase by a factor of 21.8. After it crossed the trillion yuan mark in 2007, income from land sales has maintained high levels (see Table 4.1). In recent years, income from land sales in the large cities has grown even more rapidly. In the four years from 2007 to 2010, the top ten large cities in terms of land sales income earned a total of 384.3 billion yuan, 215.8 billion yuan, 619.1 billion yuan, and 703.4 billion yuan, making an average annual increase of 22.3% (see Table 4.2).

3. Land sales income levels show huge regional differences. Due to developmental imbalance, the differences in land sales income among the eastern, central, and western regions are rather large. In 2007, when the total land sales income nationwide was 1.221672 trillion yuan, the 11 eastern provinces earned 815.33 billion yuan (66.74%), the central 10 provinces earned 231.873 billion yuan (18.98%), and the ten western provinces earned 174.502 billion yuan (14.28%). By 2009, the distribution pattern of land sales income in the eastern, central, and western regions had not changed. Total national land sales income amounted to 1.717953 trillion yuan, of which the 11 eastern provinces earned 1.258355 trillion yuan (73.25%), the central 10 provinces earned 277.269 billion yuan (16.14%), and the 12 western provinces earned 182.329 billion yuan (10.61%) (see Figure 4.1).

Table 4.2: Top ten cities in terms of land sales income, 2007–2010 (unit: billions of yuan).

Rank	City	2010	2009	2008	2007
1	Beijing	131.9	92.8	50.3	43.8
2	Hangzhou	102.5	105.4	31.2	64.6
3	Shanghai	88.0	104.3	38.2	80.3
4	Tianjin	85.3	73.2	43.9	38.8
5	Chongqing	73.3	44.0	11.7	31.4
6	Ningpo	69.3	48.8	6.8	5.2
7	Chengdu	53.5	32.4	6.6	39.6
8	Guangzhou	36.4	48.9	12.2	32.3
9	Wuhan	36.3	36.1	8.1	26.7
10	Foshan	26.9	33.2	6.8	21.6

Source: *Zhongguo fangdichan zhishu xitong shujuku* 中国房地产指数系统数据库[China Real Estate Index System Database].

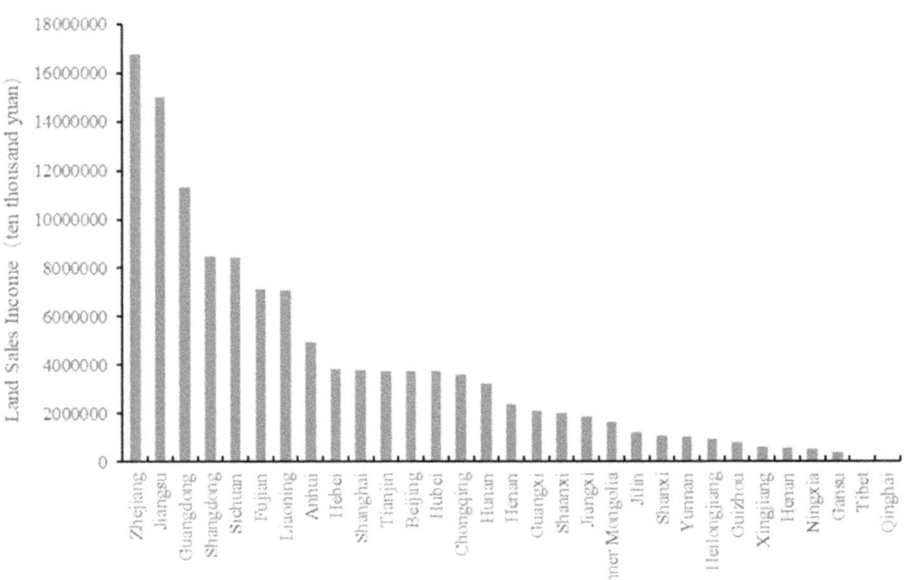

Figure 4.1: Land sales income by province (region, municipality).

1.2 Reasons for the Increase in Land Sales Income

1. The scale of land supply has substantially increased. The substantial increase in land sales income begins with the continuous expansion in the scale of land sales. In recent years, as the pace of economic development, industrialization, and urbanization has accelerated, the demand for land has grown rapidly. From 2001 to 2010, the total supply of construction-use land nationwide increased from 178,678 hectares to 432,561 hectares, that is, it increased by 253,883 hectares, or an average of 28,209 hectares per year. Of this land, the proportion of land supplied through paid use increased from 58.6% in 2001 to 68% in 2010 (see Table 4.3).

2. A swift increase in the proportion of land assigned through bid invitation, auction and listing. Among the factors contributing to the increase of land sales income, the proportion of paid sales, and especially the proportion of bid invitation, auction and listing, has been the main influence. Since 1999, China has constantly promoted reform of the system of paid use of land, reducing the proportion of land allocated for use and increasing the proportion of paid use. In

Table 4.3: Supply of state-owned construction land by payment, 2001–2010 (unit: hectares).

Year	Total supply of construction-use land	Of which:		
		Supply of land by allocation	Supply of land by payment	Proportion supplied by payment
2001年	1786,78	73,980	104,699	58.6%
2002年	235,437	88,052	147,385	62.6%
2003年	286,437	65,258	221,179	77.2%
2004年	257,920	62,054	195,866	75.9%
2005年	244,269	64,623	179,646	73.5%
2006年	306,806	63,791	243,015	79.2%
2007年	341,974	76,088	265,886	77.8%
2008年	234,185	62,381	171,804	73.4%
2009年	361,649	122,288	239,361	66.2%
2010年	432,561	138,267	294,294	68.0%

Sources: Data for 2001–2009 from Zhongguo *Guotu ziyuan tongji nianjian《中国国土资源年鉴》[China land and resources statistical yearbook]*; 2010 data from Zhongguo *Guotu ziyuan gongbao* 《中国国土资源公报》 [*Bulletin of China Land and Resources*].

particular, in May 2002, the Ministry of Land and Resources issued its no. 11 directive entitled "Regulations on the Sales of State-owned Land Use Rights by Means of Bid Tendering, Auction, and Quotation" (《招标拍卖挂牌出让国有土地使用权规定》), which stipulated that land being used for various for-profit operations in commerce, tourism, entertainment, business residences, and so on had to be assigned by means of bid tendering, auction, and quotation. Furthermore, in 2004, the State Council issued its "Decision on Deepening the Reform of Strict Land Management" (《关于深化改革严格土地管理的决定》; no. 28), which stipulated that industrial-use land must be assigned through bid tendering, auction, and quotation, so that the proportion of construction land assigned by this method rapidly increased. From 2001 to 2010, the area of land assigned use by this method as a share of total assigned land area increased from 7.3% to 88.3%. By 2010, the income from land sales via bid tendering, auction, and quotation had reached 92.23% of the total land sales income (see Table 4.4).

3. The overall level of land prices in the country has risen. Affected by factors such as rising demand for land, higher proportion of paid use of land, and higher land use costs, the national land price level continues to rise, so that an increase in land sales income becomes inevitable. From 2000 to 2010, the national comprehensive land price level and the price levels for commercial, residential, and industrial land increased, respectively, by 190.23%, 224.27%, 362.42%, and

Table 4.4: Area of land assigned use via bid tendering, auction, and quotation as a share of total assigned land area, 2001–2010.

Year	Percentage of total area (%)
2001	7.3
2002	14.6
2003	27.9
2004	29.2
2005	34.6
2006	28.5
2007	50.9
2008	83.9
2009	85.3
2010	88.3

Source: *ZhongGuo Guotu ziyuan tongji nianjian 《中国国土资源统计年鉴》 [China land and resources statistical yearbook].*

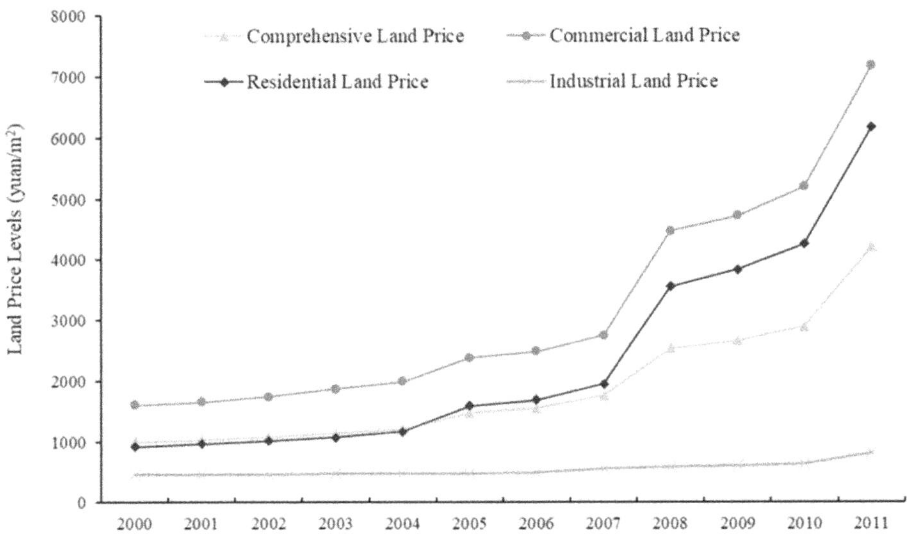

Figure 4.2: Land price levels nationwide since 2000 (yuan/m²).

39.47%. Since 2006 especially, the national comprehensive land price level has risen even faster. At the end of the fourth quarter of 2010, the overall level of land prices in the 105 major monitoring cities around the country was 2,881 yuan per square meter, an increase of 86.6% over 2006 (see Figure 4.2).

2. Costs, Benefits, and Expenditures of Land Sales

2.1 Costs of Land Sales

There are two costs of land sales: the compensatory cost from obtaining the land and the development cost paid in the initial development of the land (see Table 4.5).

1. *Compensatory costs.* These consist mainly of compensation fees for land expropriation and demolition, subsidies for land-expropriated farmers, and resettlement fees for the employees of business enterprises.

First, compensation fees for land expropriation and demolition. These fees include the expenditures by municipal and county governments in the process of land expropriation, demolition and acquisition, that is, land compensation, resettlement subsidy, supplemental compensation for objects attached to the land, young crops, and so on. From 2008 to 2010, the expenditures for land

Table 4.5: Composition of land transfer costs (units: billion yuan, %).

	2008	Percentage	2009	Percentage	2010	Percentage	Average ercentag
Land sales income	994.21	–	1,423.97	–	2,939.798	–	–
Compensation fees for land requisition, and demolition	366.213	0.368	498.567	0.35	1067.7	0.363	0.36
Subsidy to landless farmers	15.77	0.016	19.491	0.014	44.966	0.015	0.015
Initial land development	90.338	0.091	132.246	0.093	256.115	0.087	0.09
Resettlement of enterprise employees	78.241	0.079	106.648	0.075	333.663	0.113	0.089
Operational expenses of land sales	7.105	0.007	8.689	0.006	15.745	0.005	0.006
Subtotal of Costs	557.667	0.541	765.641	0.538	1,718.199	0.584	0.554

expropriation and compensation for expropriation and demolition were equivalent on average to 36% of the land sales income.

Second is the subsidy paid to land-expropriated farmers. This is a social security expense paid by municipal and county governments to land-expropriated farmers and a subsidy paid out to keep them at their original standard of living. From 2008 to 2010, this expense was equivalent to 1.5% of the land sales income.

Third is the resettlement fee for business enterprise employees. These are paid by municipal and county governments to the employees of bankrupt and restructured state-owned and collective enterprise employees. From 2008 to 2010, these costs were equivalent to 8.9% of the land sales income.

2. *Development costs.* These consist mainly of the fees paid by municipal and county governments for initial development of the land. From 2008 to 2010, this expense was equivalent to about 9% of the land sales income. In addition, operational costs relating to land sales, such as land surveys, assessments, transaction services, announcements, site rentals, bidding agents, and bid evaluations are a part of these costs. From 2008 to 2010, these operational costs were equivalent to about 0.6% of the land sales income.

Taken together, in the three years from 2008 to 2010, when land sales income nationwide, using the full numerical measure, amounted, respectively, to 994.21 billion yuan, 1.42397 trillion yuan, and 2.939798 trillion yuan, the cost expenditures were 557.667 billion yuan, 765.641 billion yuan, and 1.71181 trillion yuan, respectively. Thus the land sales costs were equivalent to 54.1%, 53.8%, and 55.4%, respectively, of the land sales income, calculated with the full numerical measure (see Table 4.5).

2.2 Land Sales Revenue

Land sales revenue is the balance of total land sales income minus the total land sales costs. It can be seen from Table 4.6 that in the three years from 2008 to 2010, the revenue from land sales amounted to 436.543 billion yuan, 658.319 billion yuan, and 1.221599 trillion yuan, respectively, which was equivalent to 45.90%, 46.20%, and 41.60% of the total land sales income for those years. The three-year average works out to 44.6%. In 2009 and 2010, land sales income exceeded expenditure, yielding balances of 140.091 billion yuan and 101.853 billion yuan, respectively. From this, we can see that only the revenue from land sales can be the real revenue that local governments get from land sales, and it represents the financial power that local governments can control. From the financial perspective, the only part of land sales income that becomes the local governments' disposable financial resources is the truly meaningful "land finance."

Table 4.6: Land sales revenue and expenditure structure (unit: billion yuan).

	2008	2009	2010
Land sales revenue	456.308	686.884	1221.599
Education expenditure	–	–	–
Farmland water conservancy	–	–	–
State Land Revenue Fund	30.891	44.063	100.789
Agricultural Land Development	12.607	14.3	18.887
Affordable Housing Projects	14.557	18.71	46.362
Urban construction	302.402	334.099	753.167
Rural infrastructure construction	33.83	43.31	102.168
Land use fees for additional land designated for construction projects	63.871	92.311	98.373
Expenditure from land sales revenue subtotal	458.158	546.793	1119.746
Land sales revenue balance	-1.85	140.091	101.853

Notes: (1) Land sales revenue is that year's land sales income minus costs and compensatory fees. (2) Education expenditure, farmland water conservancy expenditure, state land revenue fund and rural land development expenditure are specific funds; any balance, by regulation, must be passed down to the next year, so the listed figures represent the amounts that went into the funds for that year. (3) Land sales revenue balance is the difference of that year's land sales revenue and categorized expenditures on that revenue. (4) Expenditure for education and farmland water conservancy began in 2011.

2.3 Structure of Outlays from Land Sales Revenue

Based on relevant laws and regulations, the revenue from land sales is used mainly in the following eight areas (see Table 4.6 for details):

1. *Urban infrastructure construction.* There is no regulatory provision on the proportion of funds used for urban infrastructure construction. As for actual expenditure, in 2010, land revenue expenditure nationwide amounted to 979.517 billion yuan, of which 753.167 billion yuan, or 76.89%, was spent on urban infrastructure.

2. *The funds-fees for the compensated use of additional land designated for construction projects.* These are used in developing cultivated land and in land consolidation, and they are levied according to prescribed grade standards. In 2010,

the amount of land use fees for newly added construction land amounted to approximately 9% of the land sales revenue.

3. *The Agricultural Land Development Fund.* This fund, the purpose of which is self-explanatory, is accrued at not less than 15% of the revenue from land sales. The actual amount collected follows prescribed level standards. In 2010, the amount paid into the Agricultural Land Development Fund was equivalent to about 2% of the land sales revenue.

4. *The Fund for Urban Affordable Housing Projects* (城镇保障性安居工程资金). This is used for the construction of such housing, and it is to be paid at a rate not less than 10% of the revenue from land sales. In 2010, the actual expenditure for this fund was 46.362 billion yuan, which accounts for about 3% of the land sales revenue.

5. *The State Land Revenue Fund.* This is earmarked for land acquisition and reserve. The specific amount paid into it is determined by each province, district, and municipality, and it is generally about 3–5% of the land sales revenue. As it has been implemented in the past few years, though, the actual rate is about 8% of the land sales revenue.

6. *Expenditure for rural infrastructure construction.* No regulatory provision has been stipulated for this, but in 2010, such expenditure was equivalent to 10% of land sales revenue.

7. *The Farmland Water Conservancy Fund* (农田水利建设资金), accrued at 10% of land revenue. This was first implemented in 2011.

8. *The Education Fund.* This is accrued based on 10% of the land revenue, and it, too, was first implemented in 2011.

Judging from the arrangements for the expenditure of land sales revenue, most of that revenue, in the beginning, was used for urban construction, but it has been increasingly spent for the development of agriculture, villages, and people's livelihoods. The policy orientation toward the more equitable and rational distribution and use of land value-added income is obvious (see Table 4.6).

3. The Main Problems in the Management of Land Sales Income and Expenditure

Since 1989, China's policies on the distribution of land sales income has gone through three stages:

The first stage (1989–1993), characterized by shared central and local finances. After 20% of the state-owned land sales income was deducted for urban development and construction costs, 40% was turned over to the central government, and the remaining 60% was retained as revenue for local governments. The outcome of this policy was that the percentage of land sales income shared by the central government gradually decreased, from 40% in 1989 to 5% in 1992.

The second stage (1994–1998), when all land sales income went to the local government. In 1994, the fiscal tax-sharing system was implemented, and it was clearly stipulated that the income from paid use of land was to be included in the local governments' fixed income—all of it was controlled and used by the local governments. The total price of land sales was put under the exclusive fiscal management of local governments, net income was put under the management of the local state treasury, and land sales income and expenditure were under both internal and external budgetary management.

The third stage (1999 to the present), characterized by different management stages implemented for land sales income based on additional land designated for construction projects or existing construction land. To put strict controls on the occupation of cultivated land and encourage localities to revitalize existing construction land, Article 55 of the *Law of Land Administration*, revised in 1998, stipulated that 30% of the land compensation fees for new construction land would be handed over to the financial authority of the central government, and 70% would be retained by the local people's governments, to be used entirely for the development of land for cultivation.

However, there are still a few problems in the management of land sales revenue, distribution, and expenditure that urgently need to be solved.

3.1 Uneven Distribution of Land Sales Revenue Between Urban and Rural Areas

1. *Cities still use most of it.* Currently, the expenditures used mostly in cities are for urban construction, the State-owned Land Revenue Fund, and affordable housing projects. The proportion of expenditure that went for urban construction in each of the three years from 2008 to 2010 was 66.26%, 48.64%, and 61.25% (average for the three years: 58.76%) of the land sales revenue, the proportion that went to State-owned Land Revenue Fund was 6.77%, 6.41%, and 8.25% (average: 7.43%), and the proportion that went for affordable housing projects was 3.19%, 2.72%, and 3.80% (average: 3.37%). Totals for the three items amounted to 76.23%, 57.78%, and 73.80%, respectively (average: 69.56%).

2. *The land sales revenue used for rural areas and agriculture still tends to be low.* Among the expenditure items, those used for agriculture and rural areas

are primarily the Agricultural Land Development Fund and rural infrastructure construction. The proportion of land sales revenue that went to the Agricultural Land Development Fund in each of the three years from 2008 to 2010 was 2.76%, 2.08%, and 1.55%, respectively (average: 1.94%), and the proportion that went for rural infrastructure construction was 7.41%, 6.31%, and 8.36% (average: 7.57%). Totals for the two items amounted to 10.18%, 8.39%, and 9.91%, respectively (average: 9.52%).

Therefore, the current trend in the actual expenditure of land sales revenue is clearly "heavier in the cities and lighter in rural areas." Following the implementation of relevant policies, such as increasing land sales revenue support for agriculture and rural areas, accruing education funds, and farmland water conservancy infrastructure, the structure of land sales revenue and expenditure will change accordingly, and the proportion of that revenue being spent on urban construction is expected to gradually decrease.

3. *The share of land compensation for farmers from land sales income has not increased, but has fallen instead.* In the five years from 2006 to 2010, even though the compensation standards for expropriated land everywhere have generally increased by more than 30%, farmers' share in the income from land sales has not increased, but has decreased substantially. According to statistics on land expropriation compensation and land sales income in Liaoning, Zhejiang, and Henan provinces, compensation fees for land expropriation as a share of land sales income in the three provinces in 2007 accounted for 16.82%, 8.62%, and 19.58%, respectively, but in 2010, the percentages had dropped by more than a third, to 10.37%, 2.91%, and 12.99%. The increase in land sales income meant that more of the revenue from the land was being taken and dominated by the local governments. Such land revenue is attributed primarily to local governments' interest distribution mechanisms, which make them keen to expropriate and use more and more land, at low cost and high yield, to "use land to generate wealth" (*yi di sheng cai* 以地生财). This has resulted in an excess of land being expropriated, and once expropriated, it is inefficiently used. This is not conducive to protecting farmland and using land sparingly and intensively, and it seriously damages farmers' land property rights.

3.2 Problems in Land Sales Income and Expenditure

1. *Some local land sales incomes are being reduced or deferred.* Nationwide surveys have shown that the widespread problem of direct reductions of land sales income has been greatly reduced, but the problems of noncollectable levied fees,

disguised reductions, and low-cost sales are still conspicuous. These problems are manifested mainly by the practice in some localities of illegally reducing or deferring land sales income. Some governments, seeking to attract investment, will return their land sales income in the form of disguised reductions, artificially lowered sales prices, omission of collection, deferred collection, collecting fees and returning them later, using awards in lieu of investment, paying subsidy, and other forms of returning what they collected in land sales income. Some places will take profit-oriented land, which should be assigned through bid tendering, auction, and quotation, and assign it instead through agreements. In a few places there exist problems such as setting up "transition accounts" (*guo du hu* 过渡户) and using departments not responsible for financial administration, such as the local Development Zone Management Committee (开发区管委会), to handle the land sales income. Many places do not follow the requirements that the assigned land be already completely "cleared" before sales, with the result that sometimes the demolition and resettlement fees are deferred or not collected, or agent construction fees are used to defray land sales fees, or business enterprises delay payment of the land sales price.

2. *There are defects in the policy and implementation of land sales expenditure.* These appear in the following ways:

First, regulations on land sales expenditure are not clear. Since land sales costs and revenue are not clearly defined, and no definite specifications have been set on scale, quantity, and use with regard to just how much of the land sales income should be used for urban infrastructure, it is extremely easy for local governments to use the vast majority of the land sales revenue on urban infrastructure. In 2009 and 2010, the proportion of land sales expenditure that went for urban construction was 27.1% and 27.9% respectively. If costs are deducted, the proportion of land sales revenue that was used on urban construction was as high as 77.6% and 76.9%, respectively.

Second, there are defects in the policy on the compensated use fee of additional land designated for construction projects. The standards on fee collection of additional land designated for construction projects and on the accrual of the Rural Land Development Fund have not been adjusted since 2006. However, the overall level of land prices in the 105 major cities monitored throughout the country at the end of 2010 had increased by 86.6% compared with 2006, so that *the compensated use fee of additional land designated for construction projects* have obviously become too low. In addition, according to the provisions of the *Law on Land Administration, the compensated use fee of additional land designated for construction projects* are to be used entirely for the development of cultivated land, but later on, the state successively introduced the Agricultural Land Development Fund, funds for low-rent subsidized housing, and other things calling for the regular accrual or outlay of revenue *from additional land designated for construction projects*. This resulted in an overlapping of policies on how the compensatory use fees for newly added construction land would be spent.

Third, expenditure from the land sales income is not in line with regulations. In some places, the funds that are supposed to be used for urban infrastructure construction are only partly used for that purpose—they are also used to pay off government and corporate debts, build new buildings, supplement public funds, or illegally accrue debt repayment funds. Meanwhile, problems such as not allocating enough for the Agricultural Land Development Fund, the funds for low-rent subsidized housing, and the State Land Revenue Fund, or paying into them slowly, are widespread. The mismatch between the distribution and demand for special funds from land sales revenue is another conspicuous problem. For example, in places like Shanghai and Beijing, there is no farmland in their city centers, so that when they put money into the Agricultural Land Development Fund, they have nowhere to use it, so they wind up with idle funds. The funding for the Education Fund and Water Conservancy Construction Fund from land sales revenue will also encounter similar problems.

Fourth, the management of the land reserve funds is not entirely reasonable. Judging from surveys, funding arrangements from the government and from the State-owned Land Revenue Fund are inadequate, but the financial authorities can disburse funds from land sales income to defray the cost of purchasing reserves only after the land sales occurs. Thus the purchases by most land reserve institutions depend on large quantities of loans to acquire and maintain land reserves, which erodes the revenue from land. Meanwhile, many land purchase and reserve institutions have taken on the financing functions of local governments: the land reserve funds that they have acquired mainly through loans are used for stock and other investments, lending, construction of municipal facilities, and so on, and the debt repayment risk is rather high. A Ministry of Finance survey found that at the end of 2009, the loan balance of the Guangzhou Land Development Center (广州市土地开发中心) was 38.633 billion yuan, which amounted to 119% of Guangzhou's land sales income for 2009.

3. *The growth of land revenue is unsustainable.* Since 2000, land sales income experienced ten years of rapid growth. However, there were fluctuations in the land sales income during the 2000–2010 period. As the compensated land use system improved and the level of land marketization increased, the cost of land, especially the cost of acquiring existing construction land, rose sharply, to the

point that they approached or even equaled the land market price. At the same time, as the proportion of newly added construction land in the land supply decreased and the proportion of existing construction land increased, it became impossible to reverse the steady rise of land acquisition costs. Hence the room for city governments to make revenue from land will inevitably narrow.

4. Policy Recommendations on Improving the Land Sales Income System

The issue of land sales income and expenditure involve the legitimacy of land value-added income, the fairness of land income distribution, and the normative nature of land revenue and expenditure. In recent years, as land sales income as grown steadily, the state and relevant departments have introduced a series of policies to strictly control the collection of land sales income, define the scope of expenditure, and better manage the use of land sales income. However, the collection and distribution of land sales income involves fundamental problems in the land management system as well as problems in the distribution of land value-added income. Even though the reform of the land income system and land system must proceed at the same time, the reform of the land system would be so complicated and difficult that I shall not touch on it in this study. I shall make a few suggestions mainly on how to improve the land sales income distribution system.

4.1 Continuously Increase the Proportion of Land Sales Income Used for Funding Compensation for Land-Expropriated Farmers and Maintaining Their Long-Term Standard of Living

The source of land sales income is the land expropriated from farmers' collectives. Under the current system, the compensation given by local governments to land-expropriated farmers is based on a multiple of their annual income from their original use of the land. In recent years, the government has explored ways of ensuring that the living standards of land-expropriated farmers do not decline. The first way has been increasing the compensation standard for land requisition, and the second has been urging local governments to establish social security systems for land-expropriated farmers. However, with land values constantly rising and land sales income constantly climbing, these two approaches have not in any way changed the unfairness to farmers collectively in the matter of distributing income from land sales. Even though the proportion of land requisition and demolition costs have now reached 37% of the land sales income, the biggest increase in this proportion was due primarily to the large increase in the expenditures for demolition and compensation in urban-rural fringe. For most land-expropriated farmers, the compensation they receive is still quite low, the social security systems are uneven from place to place, and the level of protection is very low. To ensure that land-expropriated farmers enjoy a share of the land value-added income in the system, it is recommended that the land requisition compensation system be improved in the following ways:

1. *Establish a mechanism for adjusting compensation standards for land-expropriated farmers.* This would gradually increase the compensation standards for land expropriation as well as the share that land-expropriated farmers receive from land sales income.

2. *Effectively implement the social security system for land-expropriated farmers.* This system should be established as soon as possible, in accordance with the requirements of the State Council, in places where it has not yet been established, and land-expropriated farmers should be included in the corresponding scope of protection. Where this system has been established, safeguard mechanisms should continue to be improved and the level of protection should be reasonably determined. It should be clearly stipulated that the funds needed for the social security of land-expropriated farmers be included in the cost of land sales.

4.2 Optimize the Expenditure Structure and Guarantee a Proportion of Land Sales Income for People's Livelihoods and Rural Areas

In recent years, although the state and relevant departments have continuously introduced policies to adjust the proportion of land sales income going to cities as opposed to rural areas, rural construction, and livelihoods, the proportion of that income going to cities and construction is too high, while the proportion being used for rural areas and people's livelihoods is still too low. In the future, it will be necessary to substantially adjust the proportion of land sales income going to cities as opposed to rural areas, and to ensure the adjustment of this pattern in the system. Useful approaches would include the following:

1. *Reasonable determination of land sales cost items with respect to land sales cost expenditures.* The direct costs of land sales (compensation fees for land requisition, and demolition, expropriation and reservation expenses, and initial land development costs) and compensatory costs (social security fees for land-expropriated farmers, subsidies for maintaining the original living standards of

land-expropriated farmers, and compensatory fees for employees of bankrupt or restructured business enterprises) should be clearly identified and defined. Moreover, full payment of these costs should be ensured and prioritized.

2. *Clearly specify the proportion of land revenue to be used for urban construction, and increase the proportion of land sales revenue used for rural areas and farmers.*

3. *Coordinate the use of land sales income.* At present, the "vegetarian meat-cutting" method applied on land sales revenue has superficially increased the proportion going to rural areas and people's livelihoods, but such expedient means cannot really guarantee that the increase will in fact reach its destination. The method of constantly cutting and extending things from the land sales revenue should be changed. Instead, policies should be formulated to coordinate expenditures from land sales income, so that the system will ensure that its clear proportions are used for rural construction, farmers' social security funds, subsidized housing, water conservancy, and education.

4.3 Raise the Standards of Paid Use Fees for Newly Added Construction Land, and Apply These Funds to Establish a National Cultivated Land Protection Fund

1. *Establish a mechanism for adjusting paid use fees and land prices for newly added construction land.* This mechanism would make timely adjustments to the standards of fees collected and their scope of use, based on changes in social economic development and land prices.

2. *Establish a National Cultivated Land Protection Fund.* In view of the fact that the compensated use of additional land designated for construction projects construction land are used mainly for farmland protection and land consolidation, reclamation, and development, it is recommended that the central government set up a Cultivated Land Protection Fund for the protection of cultivated land and land remediation in key regions in order to ensure national food security.

4.4 Strengthen the Management of Land Sales Income Collection to Ensure That All Receivables Are Collected

The timely, full payment of land sales income into the local state treasuries are the basis for ensuring that land sales income and expenditure are both included in the local budget management of the funds. Local people's governments at all levels need to fully establish the system of bid invitation, auction, and listing on profit-oriented land and strictly implement the state's policy on regulating the management of land sales income and expenditure. Except where clearly stipulated by the State Council, no department should be able to arbitrarily reduce, defer, or disguise reductions on land sales income, and payments in kind of land sales income are strictly forbidden. The following methods should be adopted:

1. *Clearly specify and implement uniform management of land sales.* In agreements, bid invitation, auction, and listing sales on the use rights of state-owned land should be uniformly implemented by the competent department for state land in accordance with the requirements of urban planning, and any other units and organizations should be strictly prohibited from assigning the use rights on state-owned land.

2. *Establish strict management of land sales contracts.* In state-owned land sales contracts, lease contracts, and documents on allocation decisions, the total land sales price, rent amount, and land allocation price, as well as the payment time and method, must be clarified.

3. *Enforce the requirement that land be fully "cleared" before sales.* The practice of local governments of completing land sales before the land is fully cleared must be eliminated, since this results in the loss of state-owned land revenue. Furthermore, all kinds of platform companies should be prohibited from engaging in land acquisition and reserve.

4. *Enhance the collection of land revenue.* Establish a coordination mechanism between the state-owned land resources administration departments and branches of the People's Bank of China in order to ensure that land sales revenue is paid into local treasuries in full and on time.

5. *Improve and refine the benchmark land price system.* Municipal and county people's governments should update benchmark land prices in a timely fashion—in principle, once every three years, and in line with market changes—in order to regulate the practice of assigning land at low price.

4.5 Formulate and Improve the Management Measures for the Use of Relevant Special Funds

To strengthen the use and management of each of the special funds effectively, it will be necessary to study and formulate as soon as possible legal management measures on the use of the land sales revenue for the funds being used to invest in urban and rural infrastructure construction. The State-owned Land Revenue Fund system should be improved, with management measures on the use of that fund for the protection of cultivated land and for acquisition and reservation of land. The management of reserve funds should be improved, the scale of reserve loans limited, and the dependence on loans for reserving lands reduced. It should be further clarified that after reserved lands are realized, the local finance departments should allocate the land acquisition and reserve cost in a timely fashion to the Land Revenue Fund, to ensure that the Fund continues to grow steadily and be used.

4.6 Strengthen Supervision and Inspection and Implement Policy Strictly

A long-term mechanism for the supervision of land sales income and expenditure should be established and improved, with strengthened supervision and inspection of municipal and county-level implementation of management policy on land sales income and expenditure. The supervision should combine regular inspections with special inspections, and combine comprehensiveness with professionalism, making sure that the management policy is in fact implemented thoroughly. If an inspection turns up any behavior in violation of the regulations (e.g., the local government reduces, exempts, defers, or accepts payment in kind in land, or collects fees and then returns them, or postpones depositing land sales revenue in the treasury in a timely fashion, or does not apply the revenue to various funds), then the persons responsible will be identified, their administrative responsibilities clarified, and their punishments handled according to relevant provisions of the law. Meanwhile, efforts should be made to promote budget disclosure, increase budget transparency, report to peers in the Standing Committee of the National People's Congress, and accept supervision from society.

5 Urbanization and Land System Reform

Since the mid-1990s, China's urbanization process has greatly accelerated. From 1995 to 2010, China's urbanization rate increased at an average annual rate of 1.5%, from 29.04% to 49.95%. Urbanization has driven high economic growth, promoted industrial restructuring, and dramatically changed the production, lifestyles, and values of urban and rural residents. During this period, each percentage point increase in the urbanization rate drove a per capita GDP increase of 633 yuan. The contribution rate of urban industry to GDP increased significantly. From 1995 to 2010, the contribution rate of primary industry fell from 9.1% to 3.9%, secondary industry fell from 64.3% to 57.6%, and tertiary industry rose from 26.6% to 38.5%. Each year, more than 200 million migrating people, mostly farmers, moved to urban employment and life, which has provided unprecedented opportunities for farmers, who make up the majority of China's population, to share in the outcomes of development.[1]

During the process of rapid industrialization and urbanization in China, land has played a pivotal role. After the mid-1990s, China opened even more to the outside world, and export-oriented industrialization was implemented. To attract foreign investment and turn cheap labor into a real advantage, China has been using the arrangements of its unique land system, in which local governments, via compulsory means, expropriate land from farmers at low cost, and then they become the exclusive assigners of land use rights within their jurisdictions. By establishing park areas, providing a superb policy environment, and supplying land at low cost (and thereby avoiding the possible high land costs and land price inflation due to the scarcity of land), they ensured that in a short span of time, a decade or more, China became "the world's factory" and an important part of the global economic division of labor. After entering the new century, China's urbanization entered the fast lane. Urban development during this period benefited from a deepening of the reform of the land use system, especially with the system assigning use of profit-oriented land through "bid invitation, auction, and listing," which resulted in a substantial increase in land sales income. Land capitalization and land mortgage financing provide huge amounts of money for urban infrastructure construction.[2]

However, China's urbanization process, like its economic growth, has also shown problems of imbalance and lack of coordination and sustainability: land is being urbanized more quickly than the population is; urban development depends too much on land requisition and sales of use; the model whereby land is used to plot development has created and intensified opposition between the government and farmers, so that a huge hidden social danger now exists; and there is a huge population of migrant workers from the countryside who now live in "urban villages" and have no way of integrating into their cities, and the problem of "urban blight" is now appearing in parts of cities of all kinds. In the next ten to twenty years, China will continue to be in a period of rapid urbanization, and urbanization will continue to be one of the main forces driving China's economic growth. At the same time, China's urbanization will be in a critical period that decides whether or not the development can be healthy and sustainable; the healthy development of urbanization will to a certain extent determine the future of China's economy. Considering land's unique role in urbanization, figuring out how land will continue to play its role in promoting healthy and sustainable urbanization in the next stage of development will be crucial for the steady growth of China's economy. Reform of the land system will be key to solving problems of China's urbanization in the future.

This chapter will analyze the distinctive characteristics and problems of China's current urbanization model, especially the role that land plays in it. Through an analysis of urban renewal programs conducted in several cities in recent years, the necessity and feasibility of reforming the urbanization model are discussed. Furthermore, major reforms to the land system that some localities made in order to transform the urban development model are evaluated. Finally, policy recommendations for deepening the reform of the land system and promoting healthy and sustainable urbanization are presented.

1. Dual-Track Urbanization and Its Main Problems

In classic urbanization studies, urbanization is the epitome of human society's development, a multi-dimensional reflection of materials, space, economy, population, system, and social characteristics. It is manifested mainly as: (1) the urbanization of population, that is, an increasing proportion of the population living in cities; (2) urban growth, that is, an increasing number of cities and towns; (3) an increase in the scale of urban land use; and (4) the expansion of urban lifestyles, that is, an increase in the way urban society and behavioral characteristics influence society as a whole.[3]

However, the degree of urbanization, as measured by indicators such as urbanization rate, the number of cities, the scale of urban land use, and the pervasiveness of urban lifestyles, is the result of urbanization, not urbanization itself. If we want to understand the mechanisms that make urbanization happen and the inherent nature of urbanization, we must observe its dynamic process, especially the institutional arrangements for the population and the land when the spaces between urban and rural space are being reconfigured and the behaviors of stakeholders, and then, based on this, we must evaluate the performance and economic and social consequences of urbanization. From the perspective of urbanization dynamics, on the one hand, the population migrates from rural areas to cities, and the labor force shifts from agriculture to secondary and tertiary industries in the cities, which results in increased labor productivity, the growth of social wealth, increased income levels, and lifestyle changes. The urbanization of the population is the essence and the power source of urbanization. On the other hand, with the increased demand for land due to the population's urbanization and engagement in nonagricultural activities, land is reallocated from agriculture to nonagriculture, and cultivated land is converted to urban land, which in turn drives increases in the differential of income received for different usages of land and in the level of land capitalization. In other words, land is a catalyst for urban prosperity; land's urbanizing has become an important tool of urbanization. Since the main actors, institutional arrangements, mechanisms, and means of urbanization differ so widely, the performance outcomes of population and land urbanization and the degree to which those outcomes are shared will also differ widely.

By using this perspective to study the performance and outcomes of China's urbanization, we find that China's urbanization since the late 1990s basically proceeds on a dual track: one is government-led urbanization, and the other is the autonomous, self-initiated urbanization of farmers who make up the migrating population. The problems faced by this dual-track urbanization and the problems caused by the separation of these two tracks constitute the main problems restricting China's development of healthy and sustainable urbanization.

1.1 Government-Led Urbanization

To promote rapid urbanization, city governments at all levels in China have constantly expanded the urban circle through planning adjustments (revision of the urban master plans) and adjustments of administrative divisions (cities change into districts, counties change into cities, and rural villages change into residences), and in this way rural districts are absorbed into city maps. Under the framework in which "urban land is state-owned and rural land is collectively owned," rural collective land is transformed into urban state-owned land through the land acquisition system. As the city's scope continues to expand, the share of urban state-owned land increases, and more and more land is controlled by the government. The main features of this government-led urbanization are:

First, local governments rapidly expand urban areas through planning revisions and administrative division adjustments. We call this the "urbanization by shifting the city circle outward" (城市圈外移的城市化). From 1995 to 2010, the total urban built-up area of China as a whole increased by 20,793.8 square kilometers, or an average 1,386.3 square kilometers per year (see Table 5.1).

During the 2000–2010 decade, the development of various types of cities has shown rapid expansion from city centers to peripheral spaces. From 1999 to 2008, the built-up area of China's top ten cities expanded from 2,629 square kilometers to 7,727 square kilometers (see Figure 5.1).

Second, cities depend mostly on expropriated land for growth. With the urbanization process, urban maps continue to extend into the countryside, and the collective land of farmers is requisitioned by the government, whereupon it becomes state-owned. From 2005 to 2010, China's newly added land for residential, mining, industrial, and construction use amounted to 21,627 square kilometers. The land expropriated for these purposes amounted to 14,551 square kilometers, which means 67.3% of the total came through land expropriation (see Table 5.2).

Table 5.1: China's urban areas and urban built-up areas, 1995–2010 (unit: sq. km.).

Year	Urban area	Urban built-up area	Year	Urban area	Urban built-up area
1995	1,171,698	19,264.2	2003	399,173.2	28,308
1996	987,077.9	20,214.2	2004	394,672.5	30,406.2
1997	835,771.8	20,791.3	2005	412,819.1	32,520.7
1998	813,585.7	21,379.6	2006	166,533.5	33,659.8
1999	812,817.6	21,524.5	2007	176,065.5	35,469.7
2000	878,015	22,439.3	2008	178,110.3	36,295.3
2001	607,644.3	24,026.6	2009	175,463.61	38,107.3
2002	467,369.3	25,972.6	2010	178,691.7	40,058.0

Source: *ZhongGuo Chengshi jianshe tongji nianjian《中国城市建设统计年鉴-2008》[China Urban Construction Statistical Yearbook-2008].*

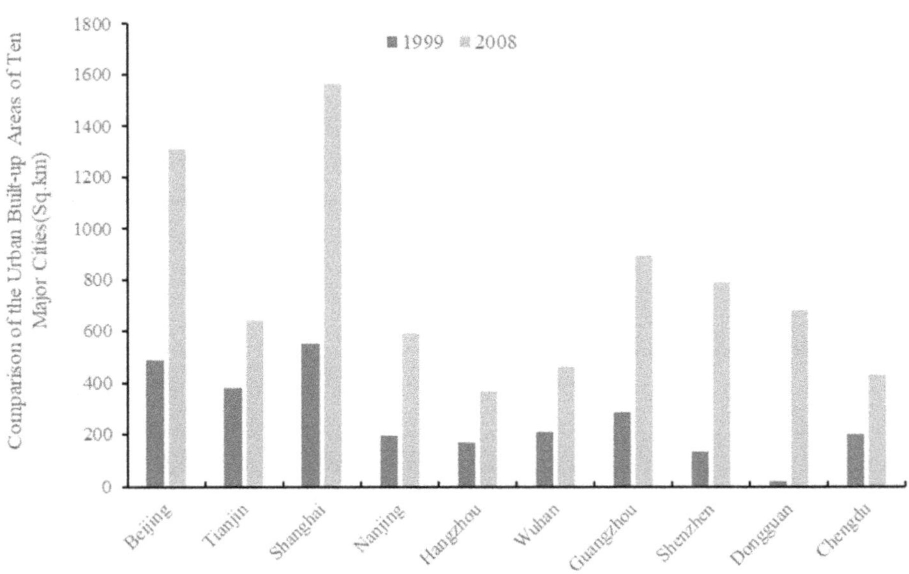

Figure 5.1: Comparison of the urban built-up areas of ten major cities: 1999 and 2008 (unit: sq. km.)

Table 5.2: Expropriated land as a share of newly added land for residential, mining, industrial, and construction use nationwide, 2005–2010 (units: 10,000 *mu*, %).

Year	Land area expropriated for residential, mining, industrial, and construction use	Additional land area designated for construction of residential, mining, industrial, and construction use	Percentage of additional land area designated for construction of residential, mining, industrial, and construction use obtained through land expropriation
2005	283.4	486	58.31
2006	329.5	588	56.04
2007	373.4	475	78.61
2008	266	445	59.78
2009	499.5	440	113.52
2010	430.9	810	53.20
总计	2,182.7	3,244	67.28

Source: *ZhongGuo Guotu ziyuan tongji nianjian*《中国国土资源统计年鉴》*[China land and resources statistical yearbook].*

Third, land sales revenue and land mortgage financing have become the main sources of funding for urban infrastructure construction. As the pace of urbanization accelerates, a huge demand for infrastructure is created. Since 2000, local governments in China have depended on land sales revenue, land reserve centers, and various financing platforms to obtain large quantities of bank loans; they use the land as collateral in order to ensure funding for land development and infrastructure construction. The funds for urban infrastructure construction come mainly from three sources: finance, loans, and self-raised capital from business enterprises. In 2008, government investment accounted for 32.2% of urban infrastructure-building funds, bank loans accounted for 29.6%, and self-raised capital from business enterprises accounted for 28.7%. All three funding sources were heavily dependent on land. The proportion of urban infrastructure investment from the central government's financial allocation dropped from 26% in the early 1980s to only 1.1% in 2008, while local governments became the mainstay of government investment. After the tax-sharing system was introduced in 1994, the local governments already established a pattern of "relying on financing for revenue, and depending on land for construction." Basically, they depended on the revenue from profit-oriented land to finance their urban infrastructure construction. Bank loans and self-financing came mostly from local city investment or financing platform companies, but the ultimate source for the funds to repay the loans remains land. Most of the bank loans are secured by pledging the land use rights and land income rights, and they are repaid through the future income derived from the land. According to the Ministry of Housing and Urban-Rural Development, the amount of land sales income that was used for urban infrastructure investment in 2009 amounted to approximately 334.1 billion yuan, which accounted for 48% of the total funding for that year (including government direct investment and subsidized investment under the name of business enterprises). Since land sales income for 2009 is calculated to be a total of 1.59 trillion yuan, 21% of the land sales fees were used for urban infrastructure construction. Let us consider investment in Chongqing as an example. From 2002 to 2008, the total financing scale there was 85 billion yuan, with bank loans accounting for 40%. Of these, land use rights and income rights allocated by the government as collateral accounted for 63% of the total loans. In addition, land consolidation revenue contributed another 13% of the total financing.

Government-led urbanization has ensured the rapid advancement of urbanization, but it has also caused a lot of problems, which are manifested primarily in the following ways:

First, land is being urbanized faster than the population is. World Bank statistics show that China's urban area has expanded at an average annual rate of 7% over the 1995–2008 period, while the urban population grew at an average annual rate of just over 3%.[4] In all of China's ten largest cities, except for Shantou, the built-up urban areas were expanding faster than the urban population (see Figure 5.2).

Once the statistical differences between China's resident population and registered population are taken into account, the characteristics of China's land urbanization rate exceeding its population urbanization rate are even more pronounced. In 2010, the urbanization rate of China's resident population reached 49.95%, but the urbanization rate based on household registration statistics was only 33.77%—a 16 percentage point difference (see Figure 5.3). Urban development has attracted a large number of migrants (mainly migrant farmers) to urban employment and life, but they are unable to enjoy the same rights as registered urban residents in the areas of social security, employment, health care, and education (see Figure 5.3).

Second, the transformation of employment structure lags behind the transformation of output value structure. In the experience of developed countries and regions, the output value and employment share of primary industry are declining, while the output value and employment share of the secondary and tertiary

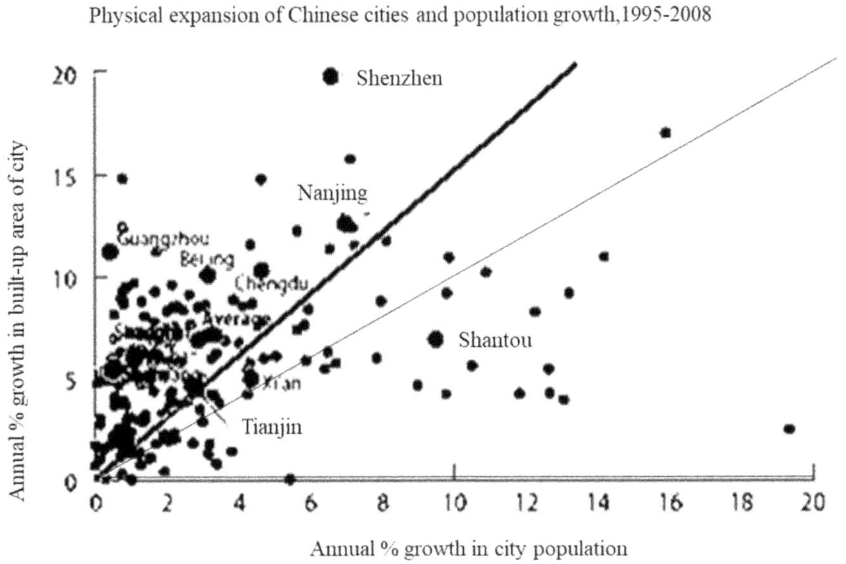

Figure 5.2: How the urban land expansion rate exceeds the urban population expansion rate.
Note: The thin line is the line of equality. The thick line is the linear regression between the two variables. The big black circles are China's ten largest cities.
Source: NBSC, various years; World Bank staff estimates.

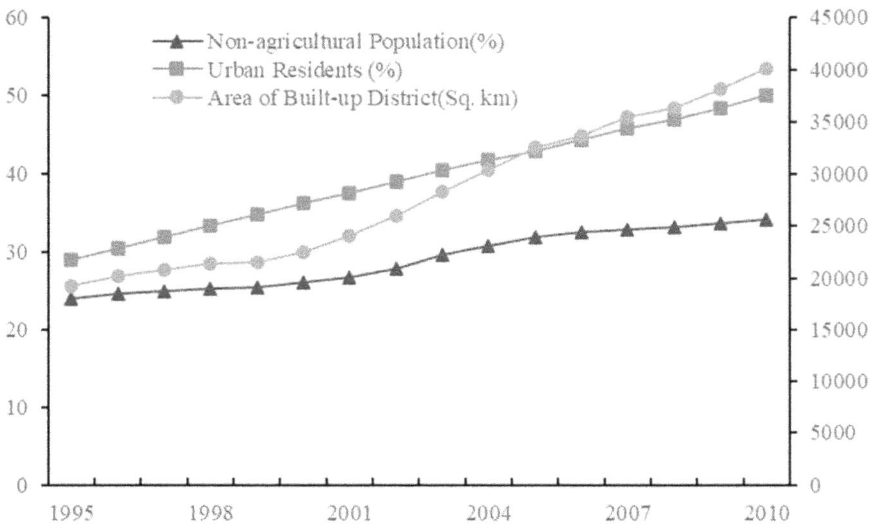

Figure 5.3: Proportion of population urbanization and urban built-up areas, 1995–2010 (units: %, sq. km.)

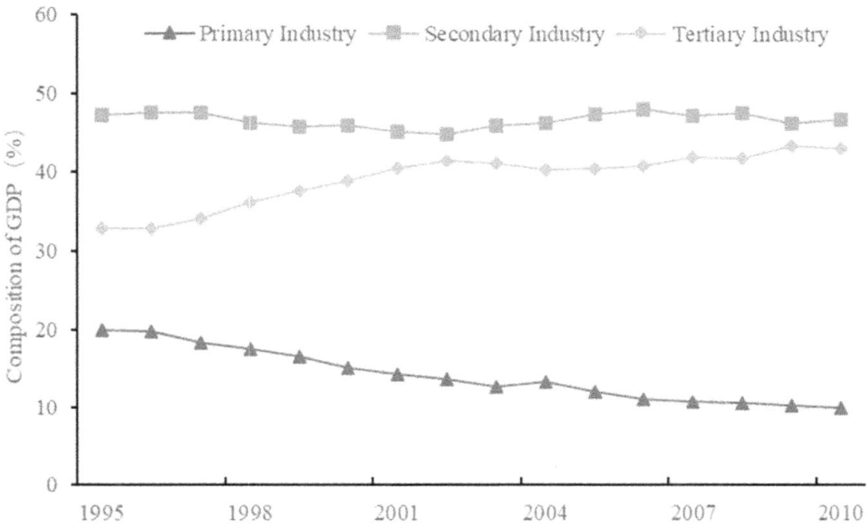

Figure 5.4: Structure of three major industrial sectors, 1995–2010 (%).

industries are rising. Under the government-led urbanization model in China, the share of the output value of primary industry in the national economy has fallen sharply, while the output value of the secondary and tertiary industries has risen sharply. However, the change in these industries' share of employment has lagged far behind the change in their output value. From 1995 to 2010, the proportion of primary industry's contribution to GNP fell from 20% to 10%, while the proportion of primary industry employment to total employment fell only from 52% to 37% (see Figures 5.4 and 5.5).

Third, the high level of dependence on land carries economic and social risks. The extra-budgetary financing by local governments is increasingly dependent on land sales revenue. Land sales revenue as a share of local governments' fiscal revenue increased from 9.1% in 2001 to 30.8% in 2009 (see Table 5.3).

The high level of dependency on land for development has brought increased economic and social risks, specifically the following:

First, financial risks for local governments. In recent years, in order to speed up urban construction and infrastructure construction, the scale of China's local government debt has increased significantly. According to the data released by the National Audit Office, by the end of 2010, the debt balance of China's local governments amounted to 10.7 trillion yuan, equivalent to approximately 26.9% of GDP that year. If we add to that the nominal indicators of public debt equivalent

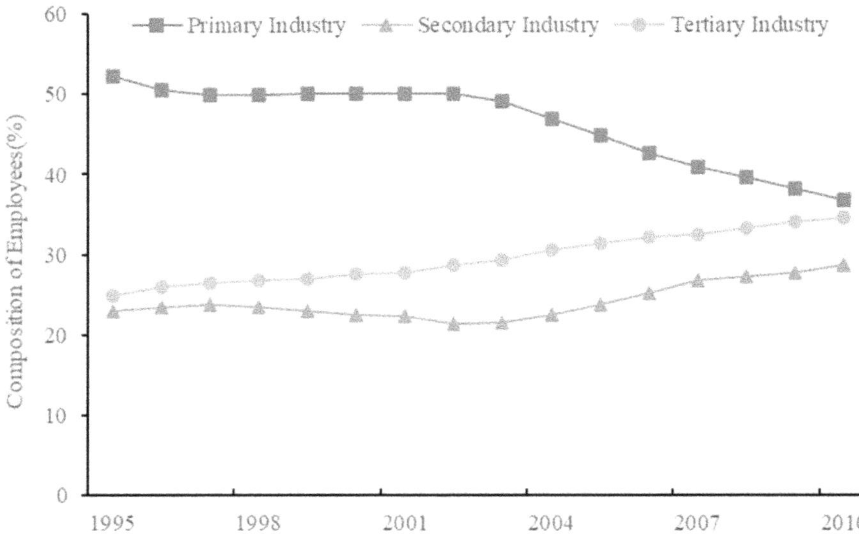

Figure 5.5: Change in employment structure nationwide, 1995–2010 (%).

Table 5.3: Land sales revenue as a share of local governments' revenue, 1995–2010（units: billion yuan）.

Year	Local government fiscal revenue	Local government extra-budgetary revenue	State-owned land sales revenue	Land sales revenue as a share of local government revenue (%)
2001	780.33	395.3	129.589	9.9
2002	851.5	403.9	241.679	16.1
2003	984.998	418.743	542.131	27.9
2004	1,189.337	434.849	641.218	28.3
2005	1,510.076	514.158	588.381	22.5
2006	1,830.358	594.077	807.764	25.0
2007	2,357.262	628.99489	1,221.672	29.0
2008	2,864.979	612.516	1,025.98	22.8
2009	3,260.259	606.264	1,717.953	30.8

数据来源：《中国统计年鉴》（2011），国土部提供数据。

Sources: *Zhongguo tongji nianjian 2011*《中国统计年鉴》（2011）[China Statistical Yearbook-2011], Ministry of Land and Resources data.

to about 20% of GDP and financial bonds issued by policy financial institution equivalent to about 6% of GDP, China's total public sector debt ratio is between 50% and 55%. It has not yet reached the 60% of GDP level, which the EU has designated as a warning line for public debt, meaning that the amount of debt is still generally controllable. However, in some regions the situation does not look optimistic. At the end of 2010, the debt ratios of local governments in the eastern and cities specifically designated in the state plan (计划单列市： these are cities that have provincial-level status in the state plan), central and western regions were 19.2%, 24.2%, and 36.4%, respectively, all of which exceed the 13–16% warning line for local governments in the United States. The central and western regions have already approached or greatly exceeded the 25% local government debt warning line used in Canada.

The main problem with Chinese local government debt is that both the debt and its servicing are highly dependent on land. Fifty percent of local governments' debt in this round relies on the local financing platforms, and land is the main collateral. Local governments' servicing of their debt depends mainly on the revenue they receive from land sales, and by repaying old debt with new debt. By the end of 2010, 37.96% of local governments' debt liability was debt liability from commitments by governments at all levels to use land sales revenue to service their debt. After 2011, the central bank raised the threshold on the use of land to accrue debt, but for most local governments, land sales revenue remains the main source of the funds they use to repay their debts. Judging from the land sales situation in 2011, the growth rate of land sales income throughout the country has fallen sharply, with land sales revenue paid into the national treasury dropping by 92.4 percentage points from the previous year. Land sales income in the eastern region has dropped especially sharply, with Shanghai, Ningbo, Xiamen, and Zhejiang (omitting Ningbo) decreasing by 22.4%, 17.2%, 17.1%, 13.4%, and 6%, respectively. The declining growth rate in land sales income has led to an increase of financial risks for local governments that have relied mainly on such income to service their debts. In the current round, 54% of the debt incurred by local governments has a loan period of five years or more. Take a look at the debt repayment year: 45.30% of the balance of local government debt at the end of 2010 will have to be paid off between 2012 and 2015—a period that represents the peak of local government debt repayment. If land sales income continues to decline in the next few years, the local government funding gap will gradually increase, and local governments' solvency and overall financial situation will be facing severe challenges.

The second is bank financial risk. Banks are the largest providers of debt funding for local governments. At the end of 2010, bank loans accounted for 79.01% of local government debt funding, and for 74.84% of it, the banks had also incurred debt with repayment obligations. If someday the decline in land sales income were to affect local governments' solvency, it would inevitably increase bank financial risk. The main debt repayment method adopted by the local financing platforms consists of "borrowing the new to pay off the old." A survey of 358 of these financial platform companies showed that at the end of 2010, the average repayment rate of these companies was 55.20%. To prevent such risks, the CBRC explicitly banned the practice of "borrowing the new to pay off the old" in 2011. However, despite this ban, the overdue debt ratio of local financing platforms is still rising. At the end of 2010, there were 148 financing platform companies with overdue debts; the overdue debt ratio averaged 16.26%, and 1,734 financing platform companies (26.37% of the total) suffered losses. In the next few years, if local governments' land sales income and land mortgage prices were to continue falling, the overdue debt ratio of the financing platform companies would further increase and the solvency of local governments would decline, at which point the negative impact on China's banking sector would not be negligible.

The third is social risk. Government-led urbanization wants only land and does not care about the people. The people flowing into the cities from outside are excluded as a matter of course. As for the farmers whose land has basically been requisitioned, after the government has given them a certain compensation and an urban subsistence allowance, they enter the city and become part of the vast army of the employed—but they, too, are excluded. Government-led urbanization has damaged farmers land rights and interests and has led to an increase of mass incidents, which have evolved into conflict between officials and ordinary people. Since the government in particular is the main entity doing the land expropriation and carrying out the demolition, it is directly causing these conflicts and making mass incidents over land expropriations more severe, more easily sparked, and more frequent. Chronologically, since 2003, the frequency of conflicts and mass incidents caused by land expropriation has increased. Among the 116 reported incidents, there was a clear increase after 2007, and they became frequent in 2010. The mass incidents that occurred during the two years 2010 and 2011 alone accounted for 62% of the reported incidents. As for their regional distribution, land expropriation conflicts are not only widely distributed but also more concentrated in economically developed areas. The 116 land expropriation conflicts were distributed across 27 provinces (municipalities, autonomous regions) and 102 counties (districts).

1.2 Farmers' Autonomous, Self-Initiated Urbanization

Through the cracks and seams of the urbanization led by local governments, there is another, more spontaneous, track of urbanization. This is the autonomous, self-initiated urbanization occurring in the various "urban-rural fringe" (*cheng xiang jie he bu* 城乡结合部) or "urban villages" (*cheng Zhong cun* 城中村), where the population is very dense and lifestyles and living environment are deteriorating.

This urbanization track is found in urban-rural fringe because, on the one hand, people from outside are flowing into the cities in search of nonagricultural employment and higher income opportunities. On the other hand, the original farmers living inside those "urban-rural fringes" that have been brought within the city circle confront the rising demand for housing the migrants and use their surplus land (including their house sites, original collective profit-oriented land, and public land) to "plant houses," which they rent out; they reap and share the benefits of differential income brought about by urbanization (see Table 5.4). Urbanization along this track presents mainly the following characteristics:

First, most of the original farmers whose land has been expropriated use their "remaining land" (the profit-oriented collective land, public land, and house sites) to build housing that they rent out or illegally transfer or sublease to persons coming to the city from outside or to business enterprises. They have thus formed a "grey market" of collective land and rental housing that is independent of the government's land and housing market. The urban-rural fringes of Beijing involve 77 street

Table 5.4: Basic conditions in major urban villages.

City	No. of urban villages	Area of urban villages (sq. km.)	Number of original inhabitants of urban villages	Number of migrants throughout city
Beijing	346	190		7,040,000
Guangzhou	138	266.48		5,000,000+
Shenzhen	320	390	5,950,000	12,000,000
Tianjin	66	8	210,000	3,800,000
Wuhan	162	213.82	356,600	1,420,000
Kunming	382	40	600,000+	1,980,000

Note: On investigation, Tianjin and Kunming clearly specified the area of housing demolished. It turned out that Tianjin had more than 8 million sq. m. of housing demolished, and Kunming had more than 40 million sq. m.; once converted into sq. km., the area is smaller.

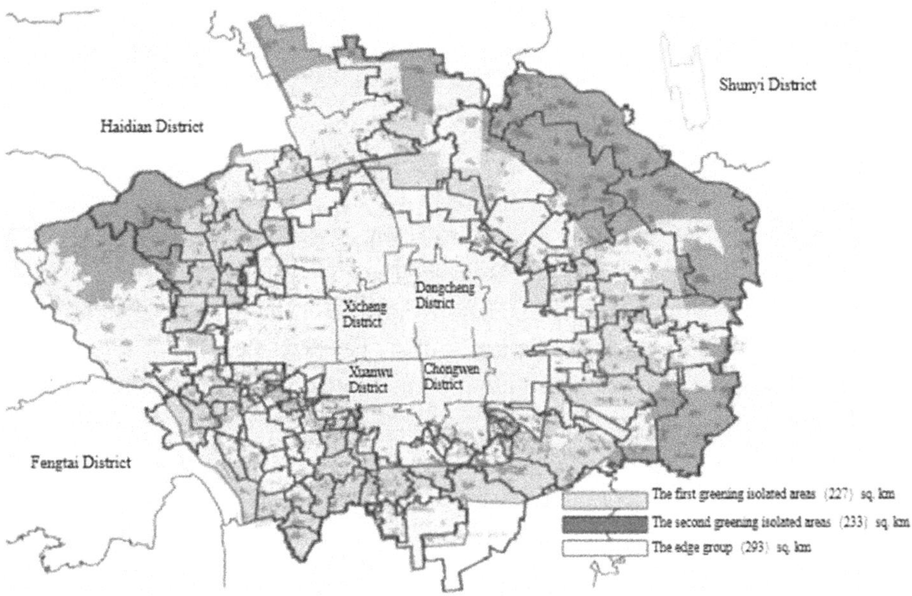

Figure 5.6: Map of distribution of natural villages in the urban-rural fringes of Beijing.

towns (街道乡镇) and 1,673 villages or communities (村, 社区), and more than 3 million *mu* of land is retained for use by farmer's collectives (see Figure 5.6).

The existing construction land area in Shenzhen is 917.77 square kilometers, and the original rural collective economic organization occupies a total of approximately 390 square kilometers, of which only 95 square kilometers is legally used (see Fig. 5.7). In Guangzhou, of the 491.1 square kilometers used by "old towns, old villages, and old factories," 356.7 square kilometers belongs to collective construction land. That is, 72.2% of the "three old kinds of land" is belongs to collectives (see Figure 5.7).

The statistics on six urban villages in Guangzhou that are rebuilding and transforming show that their total land area comprises 2,102,659 square meters; house site area totals 632,950.6 square meters, and land for economic development totals 106,062 square meters (see Table 5.5).

Second, the urban-rural fringes are the main areas for the urbanization of the floating population. Due to the low rents of houses in these areas, they have become the major gathering areas for migrants from the countryside. Up to the 1980s, the migrant population in Beijing numbered only around 200,000. By the end of 2007, however, the number of migrants living there for more than half

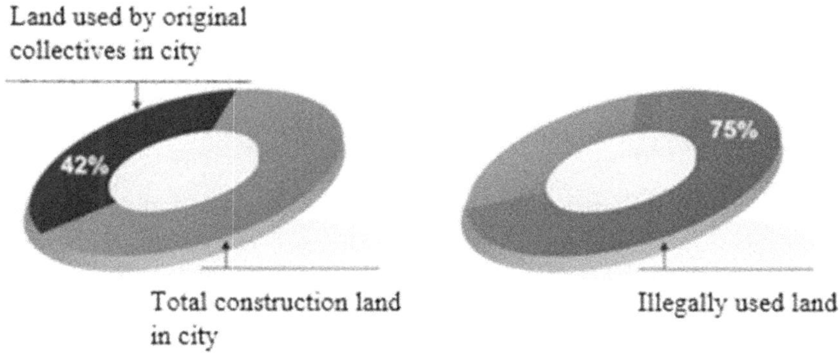

Figure 5.7: Use of the construction land of the rural collective economic organizations in Shenzhen.

Table 5.5: Land use in six urban villages in Guangzhou (unit: sq. m.).

No.	Village	Village area	Nonagricultural construction area: House site area	Nonagricultural construction area: Area used for economic development	Area used for agriculture
1	Yangji Cun	118,909	100,810	16,800	0
2	Dengfeng Cun	280,000	268,000	9,000	3,100
3	Xikeng Cun	1,500,000	115,054	13,200	13,500
4	Yaotai Cun	90,420	60,420	30,000	0
5	Siyou Cun	93,330	70,000	23,330	0
6	Nonglin Cun	20,000	18,666.6	13,732	0
Total		2,102,659	632,950.6	106,062	16,600

a year had reached 4.2 million, a net increase of 4 million from 30 years before, which accounted for 52% of the population increase. At present, the floating population in Beijing exceeds 7 million, and it grows at a rate of about 400,000 per year. The floating population living in the urban-rural fringes accounts for 52.63% of the total floating population in the city.

Third, infrastructure and public services in the urban-rural fringes are self-provided by village collective organizations. Unlike the infrastructure in areas undergoing government-led urbanization, the infrastructure in urban-rural fringes can only be self-provided by village collective organizations, since they have not been included in the urban municipal facilities systems. In addition, the public services in these urban villages, such as population control, sanitation, and public security maintenance, can only be provided by the village collective organizations, since these too are not part of the municipal urban public services systems. As a result, most of these public services in the urban villages are overloaded.

The existence of these urban-rural fringes provides local farmers opportunities to increase their income and provide differential income from urbanization by rent-sharing (see Table 5.5). At the same time, it provides convenience for those migrating from outside as they seek lower rent and living costs, and it meets the city's needs for different levels of industrial workers and service providers. However, the disorderly spread of the urban-rural fringes has also brought a large number of social issues, which eventually spawn increasingly serious "urban blight," which manifests itself in the following ways:

First, an intensified contradiction between population resources and the environment. Most villages along the urban-rural fringes of Beijing have serious population inversions, wherein the ratio between native residents and rural migrants is 1:1.2. There are 70 street towns and villages with migrant populations exceeding 10,000, and there are 667 communities (villages) where the floating population exceeds the registered population. Meanwhile, according to the statistics on the rebuilding and transformation, launched in 2010 in Guangzhou, of the "three olds" of nine urban villages covering 44 square kilometers,[5] the number of registered residents in these nine urban villages was 42,618, while the number of migrants was 67,209. The migrant-to-resident ratio was 8:5 (see Table 5.6).

Table 5.6: Demographic and economic status of six urban villages in Guangzhou (units: persons, millions of yuan).

No.	Village	Total population	Registered population	Migrant population	Average income of collectives, 2010–2012__
1	Yangji Cun	15,033	5,163	9,870	28,000
2	Dengfeng Cun	12,577	2,977	9,600	3,219
3	Xikeng Cun	32,000	12,000	20,000	1,300
4	Yaotai Cun	38,562	14,010	24,552	11,656
5	Siyou Cun	9,875	6,888	2,987	1,200
6	Nonglin Cun	1,780	1,580	200	4,000
Total		109,827	42,618	67,209	49,375

Second, a serious illegal construction problem and rising urbanization costs. To earn more compensatory income, farmers generally resort to increasing the number of attachments on the ground, which causes the land to become chaotically built-over before it is expropriated and cleared, and there is a rush of construction and planting just before clearing. In a land expropriation land requisition that took place in 1993 in Li'an Town (黎安镇), Lingshui County (陵水县), Hainan Province, there was a bizarre incident in which it seemed that 1,000 people had died—more than 1,000 tombs appeared overnight! The demolition costs of this case alone were more than 4 million yuan. When the Sanya Haitang Bay project was implemented, the local government specially set up a leadership team to handle the anticipated construction and planting rush and to carry out a comprehensive inventory, but what actually happened was that while demolition went on, other buildings were being built, and while crops were being uprooted, others were being planted—which resulted in greatly increased administrative and financial costs to the local government incurred in the land expropriation land requisition process. The demolition compensation given for some expropriated parcels where there had been a rush for planting exceeded 300,000 yuan per *mu*, more than 10 times the amount of the land compensation. In the few preceding years, Sanya Yuechuan Neighborhood Committee had built 355 illegal buildings over 150,000 square meters of land. The then-current local average for demolition cost was 2,200 yuan per square meter, so the cost of demolishing the illegal structures could not have been less than 300 million yuan. There were 19 instances of a single-family house with areas of more than 1,000 square meters; 18 cases involving buildings of six or more stories, with the tallest at eight stories; and two households that had two or more houses (see Tables 5.7 and 5.8).

According to the statistics on the six urban villages in Guangzhou, buildings in the six villages covered a total of 1,486,503 square meters. Of the latter amount, 1,284,373 square meters were licensed, and 202,130 square meters were unlicensed.

Third, the land's value is not fully realized and its use not intensive. As the city centers rapidly develop, the functions, environment, infrastructure, and industrial structure of the urban-rural fringes have been constantly improved; they become the most important spaces for the cities to grow, the places with the greatest development potential. However, at present, the industries in these areas are at the low end; housing construction standards are low, there is a lack of planning, and the farmers have no long-range expectations. This results in scattered and inefficient land development and utilization, as well as illegal land occupation, so that the full value of the land remains untapped.

Table 5.7: Illegal buildings in Yuechuan community.

Year of built	No. of buildings	Average ground area (sq. m.)	Average floor space (sq. m.)	Average no. of floors	Year of built	No. of buildings	Average ground area (sq. m.)	Average floor space (sq. m.)	Average no. of floors
1983	1	170	36.04	1	1999	5	232.6	370.57	2
1986	1	255	321.16	2	2000	10	162.8	261.98	2
1988	1	504	464.07	2	2001	7	197.14	254.87	2
1989	3	240	351.71	2.33	2002	8	208.75	376.41	2.69
1990	3	210	286.27	1.67	2003	49	195.88	409.28	2.77
1991	8	105.63	183.95	2	2004	49	159.18	350.29	2.57
1992	3	185	175.07	1.67	2005	83	182.37	531.14	3.27
1993	8	121.88	133.57	1.38	2006	53	177.11	494.91	3
1994	7	206.71	169.99	1.57	2007	17	137.18	461.95	3.38
1995	12	209.17	277.41	1.92	2008	9	169.44	515.40	3
1996	6	202.17	252.82	1.92	2009	2	271	1514	6
1997	1	150	130.20	2.50	Average		179.22	411.52	2.72
1998	9	165.22	200	1.67	Total	355	63,623.5	146,089.8	

Table 5.8: Buildings in six "urban villages" in Guangzhou (unit: sq. m.).

		Licensed			Unlicensed			
No.	Village	Subtotal	Residents	Collectives	Subtotal	Residents	Collectives	Total
1	Yangji Cun	327,169	281,609	45,560	52,630	52,630	0	379,799
2	Dengfeng Cun	304,000	110,000	194,000	90,000	68,000	22,000	394,000
3	Xikeng Cun	372,804	317,310	55,494	30,000	30,000	0	402,804
4	Yaotai Cun	92,400	92,400	0	9,000	9,000	0	101,400
5	Siyou Cun	68,000	62,000	6,000	500	500	0	68,500
6	Nonglin Cun	120,000	20,000	100,000	20,000	11,000	9,000	140,000
Total		1,284,373	883,319	401,054	202,130	171,130	31,000	1,486,503

Fourth, social problems are highlighted and intensified. Because infrastructure-building lags behind and the environmental carrying capacity is overloaded, environmental sanitation in the urban-rural fringes is poor. Litter and the flow of raw sewage is everywhere visible. Conspicuous are the problems of not enough running water and the difficulty of finding toilets. Traffic is disorderly, unsafe, and seriously congested, and in some areas unlicensed taxis operate openly. Educational and healthcare infrastructure is seriously inadequate. In particular, there are serious, hidden problems in maintaining law and order. Criminal and public security cases occur frequently, and because of the complexity and high turnover rate in the population, certain forces may be formed in some areas that cause mass incidents. In Beijing, 70% of the public security and criminal cases occur in the urban-rural fringes.

2. The Third Track: From Local Trials

In the next ten years, China will still be in a period of rapid urbanization. By 2020, the urbanization rate is expected to exceed 60%, and urbanization will play a leading role in China's economic growth. However, both government-led and the spontaneous urbanization of native farmers in the urban-rural fringes will be facing the awkward predicament of unsustainability, and it will become necessary to make a choice about where we go from here.

Let us first look at the government-led urbanization. Those who laud China's current urbanization model believe that it is precisely the benefits from "low-cost and expropriation and high-priced land sales" under the government's strong leadership that have created urban prosperity in recent years. However, in future urbanization, this institutional advantage will gradually be lost.

First, after the latest round of land expansion in some major cities, the land development intensity was already great, and this affects the livability of the space. In Guangzhou City, the land use intensity (the ratio of construction land area to state-owned land area) has already reached 22.8%, which leaves only 7.2% before it reaches the common international standard maximum of no more than 30%. Between 2000 and 2011, about 80% of the land supplied by Guangzhou City became "newly added land," with an average annual increase of 38.6 square kilometers of newly added construction land. At this rate, in less than 14 years from 2011, Guangzhou's land use intensity will exceed the 30% ecological red line for livability. In Shenzhen, according to the master plan of land use, the actual amount of land left that has potential for use in construction is only 59 square kilometers. By 2020, the area of Shenzhen's construction land will account for 49% of the city's state-owned land area. That is, it will have already passed the 30% ecological red line for livability by 19 percentage points.

Second, the cost of urbanization has risen sharply. From the perspective of either improving land use efficiency or ensuring a suitable living and ecological space, land use in most cities in the future will have to shift from building on new land to building on the existing store of land. Changing the way construction land is supplied poses a huge challenge to the existing urbanization model: when local governments try to use stock land for urban renewal and industrial upgrading, the land they use can only be the original farmers' "remaining land" in the urban-rural fringes. However, those farmers have long stopped growing crops—they have planted houses and businesses. Even though the nonagricultural activities of farmers on collective land are not protected by law, if the government compensates them on the basis of earnings from the land's original agricultural use, the process has already become fraught with difficulty. To obtain the space for urban development, it can only recognize the status quo, and as a result, the compensation for land expropriation and demolition is extremely high, which increases the cost of urbanization. From 2008 to 2010, the compensation for land expropriation and demolition in China amounted to 36% of the land sales income, mainly due to significant increases in demolition costs in the urban-rural fringes.

Now let us take a look at farmers' self-initiated urbanization. On the one hand, because of the barriers in the dual urban-rural system, even after a urban-rural fringe area is brought into a city district, it is excluded from the urban development and management system, as concretely manifested by the break-off of urban infrastructure here: the farmers' productive and living spaces are not part of any plan; farmers engaged in nonagricultural industries cannot register the companies, obtain legal permission to use collective land, or obtain urban housing guarantees. The social security, geriatric care, health care, and other basic public services that they receive are not in line with those of registered urban residents. Their management system retains its rural set up, so that they can only pay their own way and manage themselves. In other words, the self-initiated urbanization of farmers in the urban-rural fringes is basically in an "extralegal" state. Although farmers and collective organizations obtain considerable rental income by leasing out their land and housing, they have no long-term expectations, which leads them to act only for the short-term. The rural migrants who lives in these areas, not sensing that they have been fortunate, likewise do not make any long-range plans. Therefore, the self-initiated urbanization of farmers can only go one day at a time, as they exist between survival and destruction.

For China's future urbanization, it is necessary to find institutional arrangements that would integrate these two types of urbanization when they intersect, instead of letting each exclude the other. From experiments carried out in several places, we may be able to find how to blend the two.

2.1 Beijing's Experiment on Integrating Key Villages in 50 Urban-Rural Fringes in the City

To solve the urban-rural fringes' problem, increase urban development space, and enable the original farmers of urban-rural fringes to integrate into the city, the Beijing Municipal Government started an experiment in 2010 with construction in the urban-rural fringes. The area of the 50 urban villages totaled 85.3 square kilometers, equivalent to what the area of the old walled city had been for thousands of years. The 50 key villages had 200,000 original farmers, a total population of more than 1.2 million, and a population density of 14,000 (27,000) per square kilometer. The floating population in the nine districts in which the villages are located contained more than 85% of Beijing's total floating population (Figure 5.8).

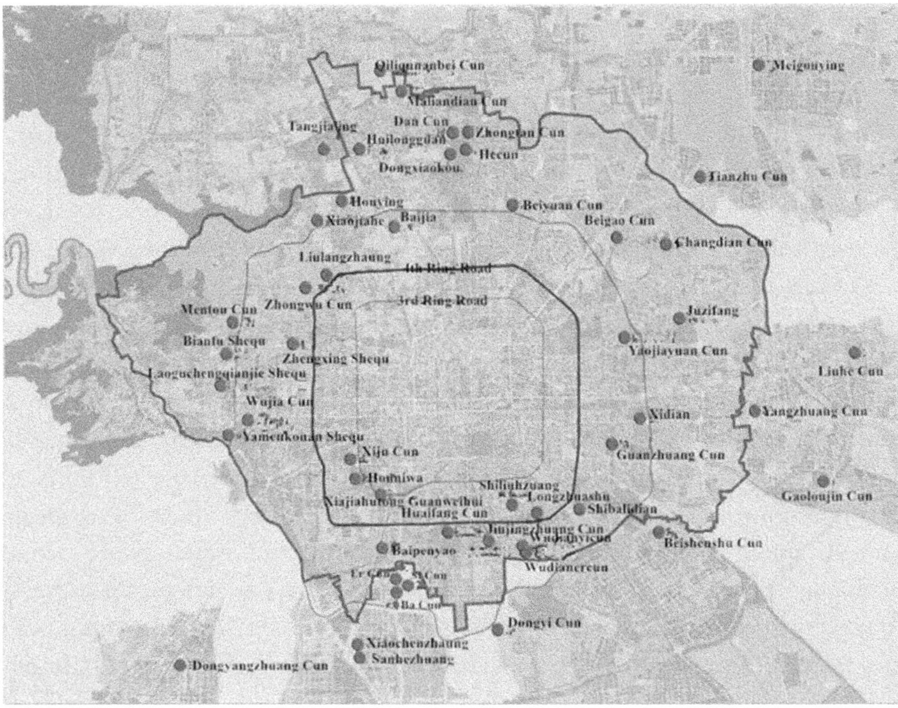

Figure 5.8: Distribution map of 50 urban villages in Beijing.

When it launched construction in the 50 key villages in urban-rural fringes, the Beijing Municipal Government made it clear that it wanted to completely change the original development methods, lifestyles, and social forms of these areas through rebuilding, but it would absolutely not use the old route of government-led urbanization with land expropriation and demolition. Instead, it wanted to ensure that while it was expanding the urban development space, the farmers would truly participate in urbanization and share the benefits of urbanization. To this end, the city set specific goals for the construction in these 50 key villages: having made the protection of farmers' interests and the complete integration of the original farmers into the city as premises, it would ensure that the farmers would become new city residents, with houses, industries, employment, shareholding, equal access to job and social security and all the basic social services of the city; it would realize the complete urbanization of the urban-rural fringes .

To make the urban-rural fringes into spaces for urban development and protect the interests of the original farmers, it is necessary to find a new urbanization model. To ensure that the urban-rural fringes integrate with their districts, Beijing made a whole set of institutional innovations that broke a hole through the dual urban-rural land system:

In planning. The 50 key villages were brought into the urban master plan, and for each village, a plan for its rebuilding was formulated and implemented. The government thus replanned the 45 square kilometers of land. Within it, 7.8 square kilometers would be used to build farmers' resettlement housing; 3.3 for farmers to develop collective industries; 13 for public green spaces to improve the urban environment; 4 for public facilities in water supply, transportation, health care, schools, and so on; and the remaining 16.9 would be put on the land market. The surplus income from the land would then be used for balancing the construction fund of the 50 key villages. Such planning adjustment and implementation will enable the 50 villages and their 45 square kilometers of land to enter the urban space system, enable the farmers in these areas to develop industry, and ensure that the construction of their housing and so on will all be included in city planning. As a consequence, they will have legal protections and guarantees that will ensure the villages' complete integration into the city.

In land policy. As the 50 key villages of Beijing are under construction, the breakthrough and innovation in land policy is the key that ensures the transformation of urban villages and the integration of farmers into the city. First, the original farmers will be allowed to reserve 50 square meters of industrial land, which they can use to develop nonagricultural industry without the collective ownership of that land having to be changed. Second, as a pilot, the Ministry of Land and Resources allowed the farmers in these villages, after their inclusion in planning, to build rental housing on their reserved collectively owned industrial land, that is, it was permitted to build housing on collective construction land. These two institutional innovations mean basically that the same rights appertain to collective construction land and state-owned construction land. Third, a certain proportion of the land in the key villages would enter the land reserve, and all of the land sales income from marketized land would be used for the municipal facilities in their areas, the resettlement housing, and funds for industrial development and social security. In other words, the funds for urbanization would be taken from the soil and used for the people. Fourth, with respect to compensation and resettlement, the compensation and resettlement for collective housing resettlement would be based on the compensation and resettlement given for state-owned land. This means that the same rights of compensation and resettlement would appertain in the resettlement of both urban and rural demolition.

In project approval. In the 50 key villages, industrial construction, infrastructure-building projects, the construction of resettlement housing and rental housing, and so on is permitted, and the rural collective economic organizations become the main entities permitted to establish their own projects and make investments. This has opened a channel for farmers' participation in urban construction. In the construction of the resettlement housing and collective industries that has already been launched, rural collective enterprises have undertaken 60% of the projects.

In social security. While completely converting agricultural land to urban residential areas, the labor force of the key villages was entirely included into the social security system according to the social security standards for urban workers, so that they could have the same level of basic public services as urban residents.

In the property rights system. In the reform of the rural collective economic property rights system, assets were quantified and people became shareholders, so that the collective economic organizations were transformed into a new kind of cooperative economic organization in which the farmers became urban residents with assets.

In social management. After the urban-rural fringes were completely rebuilt, the original rural collective economic organizations were transformed into urban communities, thus forming a unified urban-rural social management system.

After nearly two years of implementation, the reconstruction of the 50 key villages in Beijing achieved initial results. By the end of November 2011, demolition had started in all 50; in 30 it was complete, and it was underway in the other 20. Basically, there were no petitions and no mass incidents. The pilot project for the construction of rental housing on collective land had been launched. Tangjialing (唐家岭) planned to build 100,000 square meters of rental housing, and in Meigouying (梅沟营) Village, 300,000 square meters of rental housing was under construction. Construction of resettlement housing in 35 of the key villages had already started, in 12 it was completed. Each of the original farmers in the villages where the project had already started had received housing with an average of 50 square meters as well as 273,000 yuan in cash. In 13 key villages, the conversion from agricultural to residential construction had been realized, and 34 had completed restructuring, with total assets of 6.62 billion yuan and per capita equity of 115,000 yuan. Already 122,000 farmers had been included in the social security systems for urban workers. Thus the set-up of these key villages in Beijing provides a complete set of institutional arrangements for solving the problem of urban-rural fringes. Through scientifically planned use of land resources and the rational distribution of differential income from land, Beijing ensured that the farmers would have room for development of their livelihoods, and at the same time, it expanded space for the city, achieving a win-win situation for both urban and rural areas.

2.2 Nanhai, Guangdong: Industrialization and Urbanization on Collective Land

Since reform and opening up, the unique industrialization and urbanization model developed in Nanhai[6] (南海) has enabled farmers with land to participate in industrialization and urbanization and the common development of urban and rural areas on the basis of rural prosperity.

1. Rapid industrialization on collective land. In the early 1990s, local governments at all levels and farmers' collectives saw opportunity in the large number of foreign and domestic business enterprises that were coming to Nanhai to look for land on which to build their factories, so they took the land under their control and put it under uniform planning and integrated it, whereupon they leased the land out to those enterprises. By 2002, Nanhai City's industrial land totaled 150,000 mu, of which 73,000 mu, or nearly half, was still under collective ownership. Thanks to the conversion of rural collective land away from agriculture, rural industrialization in Nanhai advanced by leaps and bounds, and one after another factory had grown on collective land.

Because Nanhai's industrialization had been initiated through the leasing of collective land in Nanhai, it avoided land expropriation and conversion to state-owned land, so that the village retained the differential income created by the direct entry of collective construction land into nonagricultural use. Since the farmers' collectives retained the collective ownership of the land, they could enjoy the differential income from that land for a long time. By 2009, the collective assets of both levels of rural collectives amounted to 23.712 billion yuan, and the disposal income of both levels of rural collectives reached 4.327 billion yuan. From 1994 to 2000, farmers' per capita net income from dividends in Nanhai increased from 1,062 yuan per year to 1.951 yuan per year, an increase of 10.7% per year. The collective land income assumed the function of providing community public goods and public services. From 1992 to 2009, the total investment in rural public goods by both levels of village collectives amounted to 769.567 million yuan, and the amount invested per year in public utilities increased from 157 million yuan to 3.330 billion yuan. The amount of investment going to public utilities as a share of total public goods investment by both levels of village collectives increased from 31.46% to 43.28%.

2. *"Developing" cities on collective land.* Around 2000, in order to promote urbanization in Nanhai City, the local government promoted the comprehensive transformation of old towns, old factories, and old villages, and it built a community of shared interests of the government, collectives, and farmers through innovations to the land control and interest distribution system. This protected farmers' interests, facilitated the effective and intensive use of land, and promoted industrial upgrading and urbanization.

First, the local government clearly defined the boundaries of historical construction land and defined the rights appertaining to them, and it provided protections for the legal use of and transactions involving collective construction land. With regard to collective land that had already become used for construction before January 1, 1999, without going through the process to make the construction legal, the collective was only required to have obtained a collective landownership certificate for it to be classified as land of construction nature. For that land, plus land that fell within the scope of construction land in city plans, land that already appeared to have been built over, and land that had already appeared through aerial imaging to have been developed—the local government defined the rights appertaining to all, and granted permission for them to be rebuilt.

Second, the local government developed guidelines on transactions involving differently tenured land. In situations where the retained land is state-owned, if the land use rights-holder does not change and he changes the land's use, then the original land user may rebuild on the land, with the land exempt from public trading. Situations in which the land, after rebuilding, is used for developing commercial real estate are handled in accordance with the provisions on public transactions of land. When rural collective construction land is leased out, assigned, transferred, mortgaged, or circulated, it must be done in accordance with the *Guangdong Province Implementation Measures on the Transfer of Use Rights of Rural Collective Construction Land* (《广东省农村集体建设用地使用权流转实施办法》) and similar regulations of Foshan City and its various districts. If the rural collective construction land involved needs to be changed for commercial real estate development, then, as required, the land is transferred to state ownership and public transaction procedures are followed.

Third, the local government formulated measures for the transfer of collective construction land to state construction land. Collective construction land can be replaced by other land within the collective. When collective construction land within an industrial park is converted to state ownership, a land use approval document is issued, which specifies the nature and the area of the land, and the legal procedures are followed to convert that land to state ownership. If a building involved simultaneously occupies state-owned and collectively owned land, the part under collective ownership is converted to state ownership provided that the rural collective economic organization voluntarily consents to it. When collective land is converted to state ownership and publicly assigned, all of the original persons of the collective are allowed to participate in the allotment and shareholding of the completed property, in order to protect the long-term income of the original collective members. Commodity factories converted from old factory buildings may be transferred, sold, or leased out in accordance with the relevant regulations. Units and individuals that purchase commodity factories may apply for the relevant property rights certificates in accordance with the relevant regulations.

Fourth, the local government formulated a land sales fee collection policy for collective land transactions. If an old factory that was originally on rural collective land is going to be rebuilt, and the land is only converted to state ownership but no transfer occurs, then land sales fees are not collected on it. If the land is converted to state ownership and transfer occurs, then, after the government collects the land sales fees according to the regulations and turns them over to the treasury, some or all of the special funds will be used to cover the expenditures for the living standard subsidy of the original land-expropriated rural residents, the expenditures for the social security of the original land-expropriated rural residents, the expenditure for the Agricultural Land Development Fund, and the Rural Infrastructure construction Fund.

The fifth relates to the registration of real estate ownership of farmers' apartments. In the project for rebuilding the "three olds," whenever the use rights to an original house site have been recalled (the house is demolished and the original real estate title certificate is canceled), or if a villager (resident) did not have an allocated house site on which to build a house, then to participate in the distribution of farmers' apartments, the relevant provisions for self-built houses apply for the real estate rights registration procedures. However, the farmers' apartments may not be traded on the market.

The transformation of the "three olds," which is at the core of the innovations to the land system, has released a huge amount of space for construction for the latest round of development in Nanhai. From 2007 to the present, Nanhai District has completed 118 renovation projects over 4,054 *mu* of land with an investment of about 13.2 billion yuan. A total of 779 cases of land use rights conversions were completed, covering an area of 8,341 *mu*; 1,849 *mu* of land was publicly assigned, with land sales fees totaling 6.71 billion yuan; the cost of reimbursed land sales was nearly 2.95 billion yuan; reimbursements of land sales fees totaled 10.996 million yuan, and a total of 28.903 million yuan went for demolition compensation. In this urban renewal, the conversion of land promoted industrial upgrading, and the amount of rent collected by farmers' collectives land increased substantially. This created a new, win-win situation wherein urbanization involving land-use conversion was promoted and farmers' incomes increased.

2.3 Reformed Methods of Land Requisition and Demolition in Lingshui and Sanya, Hainan

In consideration of the shortcomings of government-led land expropriation and urbanization and the conspicuous problems with land expropriation and demolition as it is currently done, Lingshui County (陵水县) and Sanya City (三亚市) in Hainan Province have explored a set of new approaches to the process, summed up in the phrase, "led by government, with farmers' participation, shared benefits, and everyone winning," in order to achieve the goals of economic development and social harmony:

Letting farmers become direct participants in land expropriation and demolition. From the start of the land expropriation and demolition process, the principle status of the land-expropriated farmers is clearly defined, and the government provides guidance, supervision, propaganda, and coordination. The village collectives establish joint-stock companies to expropriate the land and demolish housing, and the government delivers the land expropriation compensation fees and work funds as a "package" to those companies. After the companies pay compensation to the villagers, the balance is shared by all of the villagers.

Allowing farmers' collective construction land into the market. In a pilot project, Lingshui County, in addition to paying farmers' compensation according to the highest standard in Hainan Province, permitted village committees during the rebuilding of old villages to put leftover collective construction land direction directly on the market through the bid tendering, auction, and quotation system. In this way, it became the first place in China to realize land revenue by putting collective construction land through the tangible market.

Innovate by retaining land for resettlement and cultivating industries that would be suitable for farmers to have a stable income stream. To effectively relieve the worries of farmers losing their land, Lingshui County left the land with the most complete infrastructure and best geographical location in the planned project area to the collective use of farmers, in accordance with the 8% proportion of the total area of expropriated land, which would support the development of the collectives' distinctive businesses.

After the new land expropriation and demolition model was developed, the farmers' collectives received a sum of money from the expropriation and demolition and organized their own evaluation and assessment of superfices, thus saving nearly 1 million yuan in assessment costs and more than 57 million yuan in demolition compensation costs. Add to that the 10% in work expenses and 10% in unforeseen expenses, and more than 97 million yuan of development funds were received by the village collectives. After the 289 mu of collective construction land in Dadun Village were listed on the market, the village collective received 610 million yuan in land premiums. In addition, Dadun Village (大墩村) acquired 248 *mu* of reserved land, which the village committee planned to use to jointly build hotels with other enterprises. They also plan to build four hotels with 320 rooms generating public income, as well as supermarkets and markets in the new resettlement area. These may be self-operated by the collectives or leased out, but either increases income levels among the masses. Under the new approach to land expropriation and demolition, the number of complaints from Yuechuan Village (月川村), which had developed a reputation for long, dragged-out appeals and conflicts, suddenly dropped. "Zero appeals" were achieved in Dadun Village, which had waged a life-and-death resistance against land expropriation. When the farmers participated in urban construction, and when the rights appertaining to collective land, and the price, were at parity with state-owned land, the farmers' long-term livelihoods were truly guaranteed, and their income levels and living conditions underwent tremendous changes.

From the three cases above, it is evident that China can continue to urbanize in the future, as long as the current urbanization model is changed. The core of the new urbanization model would be to make a breakthrough in the dual system and allow farmers to participate in urban construction. Reforming the land system will ensure that the same rights appertain to both collective- and state-owned construction land, and it enables rural collective land in conformance with planning to be a part of the urbanization process.

3. Policy Recommendations

During the next ten years, China will still be at the rapid industrialization and urbanization stage. As the two processes accelerate, land will also enter a peak period of conversion away from agricultural use, and land supply and demand will be extremely unbalanced. There will be a high risk of social conflicts over land occurring, so proper handling of various land use relationships—especially the relationships between people and land, land and other resources, and person-to-person relationships—will promote social stability and harmony. We must reform the land system, fundamentally change the patter of development driven on land, and promote healthy and sustainable urbanization.

3.1 Reform of Property-Based Market Allocation Methods

1. *Equal protection of land property rights.* A law-based system centered on protecting farmers' land property rights should be constructed. The central government should uniformly ensure that all land resources are confirmed, registered, and certified; it should confirm the land rights relationships between landowners and land users; and it should provide basic institutional services for safeguarding land property rights and facilitating land rights transactions. The situation wherein different rights may appertain to the same parcel of land if it is under two different ownership systems should be changed, so that the same rights to tenure, use, revenue, and disposal appertain to land whether it is collectively owned or state-owned; this would be the realization of "same land, equal rights" under the protections of the *Constitution* and relevant laws. Fair compensation for the land property rights of farmers who lose their land should be gradually implemented, with the ultimate goal being compensation at the market price of expropriated land. Local governments should basically be oriented toward compensating farmers who lose land and property and be encouraged to adopt various means, such as reserving land, replacing property and businesses, increasing compensation standards, for farmers and collectives that lose their land. The rights of land-expropriated farmers to speak and negotiate when their land is being expropriated should be respected and protected.

2. *Confirm equal access rights to land under different types of ownership.* Use controls should be the only access system. Under use controls, farmers' collectives should have the same right to bring their land into nonagricultural use as other landowners do and to enjoy the same kind of nonagricultural value-added income from that land. Since the main purpose of clear planning is to bring into reality a functional spatial layout, the pattern wherein local governments use planning to transform farmers' collectively owned land into state-owned land and government operation should be changed. We should break the current policy in which

land under different types of ownership, depending on whether it falls inside or outside a boundary separating a city from a village, determines access. Apart from land outside the city circle that is used for nonagricultural construction not in the public interest, it should also be possible for land owned by farmers' collectives within the city circle to be used for nonagricultural construction without the land having to change ownership, as long as that construction is in compliance with use controls. It should be clearly specified that urban land reserved for a construction zone be stated-owned, and that newly added construction land used for nonagricultural construction may be retained under collective ownership, except where it is requisitioned for an item of public interest. As for collectively owned land in built-up areas, it should be possible to protect farmers' land property rights by means of "protecting rights and distributing the interests" or "transferring rights and protecting the interests."

3. *Reconstruct the market in which land is traded equally.* An open, fair, equitable, and unified trading platform and trading rules should be established in order to achieve a land market where there is equality in ownership and where market transactions determine supply and demand and prices. The current arrangement in which local governments are the exclusive, monopolistic suppliers of land should be broken, in order to set up a uniform trading platform in which land under different types of ownership are supplied on an equal basis. As long as relevant laws and trading rules are complied with, then regardless of who the traders are—whether the government, farmers' collectives, users of state-owned land—they can all engage in land trading on the uniform land transaction market. Second-tier land markets should also be invigorated, that is, we should promote development and refinement of markets for land mortgage, rental, and use sales.

4. *Speed up the process of "dual-track merging into single-track" of land supply and reduce the system of nonmarket allocation of land resources.* The market allocation reform of state-owned land should be further expanded, while the scope of allocated land use should be gradually shrunk. The coverage of paid land use should be steadily expanded, until the "dual-track system" of land supply is finally removed. Paid use should be implemented first for profit-oriented land that is competitive, generates operating income, is ripe for reform, has municipal facilities, and has all kinds of social enterprise. In the near-term future, except in the case of special land—military land, affordable housing, land used for state security and maintenance of public order, etc.—for which the allocation system may continue to be used, all other kinds of construction land should be allocated through the market.

3.2 Fundamental Reform of Land Finance

1. *Distribute the government's profit-oriented management of the land.* With reference to the experience of state-owned enterprise reform, we may consider setting up a State-owned Land Asset Administration Commission, similar to the State-owned Assets Supervision and Administration Commission of the State Council (国务院国有资产监督管理委员会), that would be the government agency responsible for organizing and leading the work of managing state-owned land. At the same time, however, state-owned land companies and other economic organization should be set up, and then state-owned land that needs to be managed for profit should be delivered to them, for putting in market operations, transferring, leasing out, put under joint ventures, or used in shareholding; they would have responsibility for preserving and increasing value, and land revenue would be paid to state finance. These economic organization, as market entities, would accept governmental regulation and supervision in accordance with the uniform market rules. As state-owned companies, they would also accept the management and guidance of the newly established State-owned Land Asset Administration Commission. When necessary, it would also have to obey the requirements of the state and assume certain responsibilities for using land to serve the requirements of macroeconomic regulation and control.

2. *Clarify the government's land management functions.* After the function of profit-oriented management of land has been removed, the land management functions to be exercised by the land management departments at all levels should be further clarified in greater detail. Their main functions should be strict implementation of land use planning; strengthened land use control; effective protection of farmland; enhanced institutional infrastructure construction; improved identification, certification, and registration of different land resources; and ensuring equal protection of all types of land property rights. They should build the land trading platform, formulate the trading rules, standardize trading behavior, implement land price management, and build a unified urban and rural land market. They should enhance the lawful management of land and increase the intensity of both land supervision and the punishment of illegalities.

3. *Improve the distribution of land sales income and the management of incomes and expenditures.* The proportion of land sales income used for compensating land-expropriated farmers and guaranteeing the maintenance of their living standards over the long term should be continuously increased. A mechanism for adjusting the compensation standards for land-expropriated farmers should be established, and those compensation standards should be gradually increased, as

well as the land-expropriated farmers' share of the land sales income. The social security system for land-expropriated farmers should be effectively implemented, with an optimized expenditure structure ensuring that the proportion of land sales income is being spent for villages and farmers' livelihoods. It will be necessary to greatly adjust the proportion of land sales revenue going to cities versus villages and for construction versus livelihoods, and the adjustment of this pattern should be ensured systematically. With regard to land sales cost expenditures, cost expenditure items should be reasonably determined. The proportion of land revenue used for urban construction should be clarified, and the proportion used for villages and farmers should be increased.

4. *Standardize the management and supervision of state-owned land sales revenue.* It should be made clear that the revenue from profit-oriented management of state-owned land must not be considered as current surplus in government finances. The revenue from the use sales of state-owned land to be obtained from the specialized agency in charge of the profit-oriented management of land should be paid into a special account separate from the government's finances and uniformly managed by a separate specialized agency. The use of land sales revenue should be subject to the approval of the People's Congress at the relevant level. The government finance departments can prepare an annual budget for land revenue and report it to the People's Congress at that level for review. Land sales revenue, preparatory to payment into the government budget, must be strictly controlled, and handed over to the government after approval by the People's Congress. Expenditure items from the land sales revenue and how the fees are being used should also be reviewed by the local People's Congresses to ensure that the funds are being used longer in a more transparent and more reasonable manner.

3.3 Improving the Property Tax System by Putting It in Line with Modern Society

We should draw on the experience of market economies to establish a new land real estate tax system that befits our national resource conditions. The tax revenue system with regard to land holding, circulation, and revenue linkages should be built integrally and systematically reformed. As for land holding, the main tax levy should be changed from quantitative one to an entirely ad valorem one in order to improve tax flexibility and fulfill the function of a land real estate tax that raises local government revenue. As for land circulation, the overall tax rate, as well as the turnover tax rate, should be reduced, in order to better allocate resources and improve efficiency. As for revenue derived from the circulation of land, tax collection and management should be strengthened, in order to fully promote the fair distribution of social wealth.

3.4 Establishing a Fair and Shared Distribution System for Value-Added Income

The reform of the land expropriation system should be accelerated. To further improve the land-expropriation compensation and resettlement system, broader approaches to resettlement should be found, and various methods of reserving land for resettlement and introducing land shareholding should be explored, so that whether farmers retain or transfer their use rights, they benefit and their long-term livelihood is assured. Besides improving the dispute resolution system for land expropriation-related disputes, relief channels should be cleared so as to safeguard farmers' legitimate rights and interests on their land. Second, a clear distinction should be drawn between "returning price increases to the public" and "returning price increases to the government." That is, when the increased added-value of land "returns to the public," it should not mean "return to the local government," or even become some local government officials' unofficial income. When increased prices are returned to the public, there must be the assurance that the added-value of land is indeed given back to society. Third, a land fund system should be established. A certain proportion of land revenue should be collected and used for adjustments for dearth and surplus, to damp down the effects of market volatility on local finance, and to realize a rational, "interannual" (*nianji* 年际) distribution of land revenue.

3.5 Establishing a Unified Registration System That Serves Modern Social Management

A land registration system is one in which the state registers land in accordance with procedures and requirements stipulated by laws and regulations in order to protect the legitimate rights and interests of land rights holders and to promote land use. It is a specific means of the principle of publicity of land rights changes, and it also a legal means for establishing normal land transaction order and ensuring the security of transactions. In accordance with the requirements of the *Property Law*, laws and regulations on the unified registration of real estate (with land as the basis) should be promulgated and implemented as soon as possible, with a unified legal basis, registration authority, ownership certificates, etc.

Notes

1 See the 2010 annual report from the China Development Research Foundation (中国发展基金会), *Yi ren wei ben de chengshihua* 《以人为本的城市化》[*Human-centered Urbanization*].

2 Tudi ziyuan zhanlüe keti zu 土地资源战略课题组, ed. *Tudi ziyuan zhanlüe yanjiu* 《土地资源战略研究》 [*Study on Land Resource Strategy*] (2011), Internal Research Report

3 Paul L. Knox and Linda M. McCarthy, eds., *Chengshihua* 《城市化》 [*Urbanization*], trans. by Gu Chaolin 顾朝林 and Tang Peiyuan 汤培源 (Beijing: Kexue chubanshe, 2008)

4 Guowuyuan fazhan yanjiu zhongxin 国务院发展研究中心 [Development Research Center of the State Council] and Shijie yinhang 世界银行 [World Bank], eds., *2030 nian de Zhongguo* 《2030年的中国: 建设现代、和谐、有创造力的高收入社会》 [China 2030: Building a Modern, Harmonious, and Creative High-Income Society] (Beijing: china financial and economical publishing house, 2013), page 34.

5 The urban villages were: Pazhou (琶洲), Liede (猎德), Xiancun (冼村), Linhe (林和), Yangji (杨箕), Xiaoxintang (小新塘), Tangxia (棠下), and Sanyuanli (三元里).

6 *Translator's note:* Nanhai's administrative status has changed through the years. Nanhai County (*xian* 县) was upgraded to a county-level city (*xianji shi* 县级市) in September 1992, and then became a District (*qu* 区) of Foshan City (佛山市) in December 2002.

6

Bidding Farewell to "Development Driven by Land"

During the last strategic opportunity period, China used its unique land system to drive high economic growth and rapid industrialization and urbanization. The process was manifested in the following specific ways. First, the ample supply of land guaranteed high economic growth. From 2003 to 2012, the total annual supply of state-owned construction land increased from 286,400 hectares to 690,400 hectares, at an average annual growth rate of 10.27%. Second, the low cost of industrial land guaranteed the rapid advancement of industrialization. From 2000 to 2012, the average annual increases in the national comprehensive land price level, commercial land price level, and residential land price level were 10.04%, 11.40%, and 14.42%, respectively, but for industrial land the price level increased by an average of only 3.35% per year—far lower than the commercial and residential land price increases. Third, urban land expansion and land capitalization have promoted urbanization. Between 2000 and 2012, China's urban built-up area more than doubled, and land sales income has increased by a factor of more than 45. In 2012, land mortgage loans in 84 key cities reached 5.95 trillion yuan, making it an important source of funding for urban infrastructure construction.

However, the model of "development driven by land" (*yi di mou fa zhan*以地谋发展) has also exposed many problems, especially after land sales income dropped off in 2012, and its unsustainability and risks were exposed. Finding a source to power the next round of development through reform has become an urgent matter.

1. The Land Sales Model: Frustrated in the East, Spreading in the West and Center

1.1 The Predicament of the Land Sales Model in the Eastern Coastal Region

First, land sales income in the East has dropped significantly. In 2012, land sales income nationwide decreased by 13.7% from the previous year, and in the East it decreased by 20.6%. As for cities in 2012, even though the East had the most cities (37) among the country's top 50 cities in terms of land sales income, and the most (6) among the top 10 (see Table 6.2), the land sales income of the four first-tier cities of Beijing, Shanghai, Guangzhou, and Shenzhen decreased by 38.60%, 23.10%, 22.02%, and 50.13%, respectively from the previous year. In addition, among the top 50 cities, of the 18 cities that posted a year-on-year increase in land sales income, only 8 were in the East. Of the 20 cities with the largest year-on-year decreases, 18 were eastern cities.

Second, the growth of land area sold in the East has slowed down. In recent years, land-dependent growth in the East has slowed down. From 2000 to 2007, the average annual increase of land area sold was 25.2%, and from 2008 to 2012, the average annual increase fell back to 9.06%. In 2012, 122,700 hectares of land was sold, representing a decrease of 21% from the previous year. The proportion of land supply in the three major regions, the East, Center, and West, was 52:25:23 in 2008, but in 2012 it had changed to 38:27:35. The proportion of land supply in the East declined for four consecutive years.

Third, land prices in the East have risen rigidly, bidding farewell to the era of low-cost land supply. From 2008 to 2012, the unit price of land sales (total price of land sold / total area of land sold) in eastern China increased each year by 10.65% on average, to 3.93 million yuan per hectare. In 2009, the unit price of land sales in the East started to exceed 10 million yuan per hectare, and by 2012, despite the overall downturn in the land market, the unit price of land sales there was still as high as 11.82 million yuan per hectare. Among all types of land, the cost of industrial land rose particularly sharply. From 2010 to 2012, the average annual increase in industrial land prices in the Yangtze River Delta, Bohai Rim, and Pearl River Delta reached 8.75%, 7.19%, and 10.53%, respectively. It has already become difficult to rely on government-subsidized land prices to maintain low-cost industrialization.

1.2 The Land Sales Model Is Replicating and Spreading in the Central and Western Regions

In recent years, industrialization and urbanization in the central and western regions have accelerated, but the replication of the "development by land" model in the East is quite obvious.

First, the growth in the rate of land sales income in the central and western regions has accelerated. Before 2010, the increase of land sales income in the East was always much higher than in the Center and the West. In 2009, for example, land sales income in the East increased by 83.89%, while land sales in the Center and West increased by 25.84% and 44.92%, respectively. In 2010, the growth rate of land sales income in the Center and West began to exceed that of the East, having reached 102.24% and 145.44%, respectively, while in the East it grew only 53.19%. In 2011, land sales income in the Center and West increased by 33.6% and 40.1%, respectively, from 2010, in stark contrast to the increase of only 1.9% in the East. In 2012, with the land market in an overall decline, land sales income in the Center and West fell by 5.9% and 1.3% from 2011, but that is far less than the 20.7% decline in the East (see Table 6.1).

Table 6.1: Land sales income and growth in the east, center, and west (units: billion yuan, %).

	2009		2010		2011		2012	
	Income	Increase	Income	Increase	Income	Increase	Income	Increase
East	1258.3	83.9%	1927.6	53.2%	1964.2	1.9%	1557.9	-20.7%
Center	277.2	25.8%	566.2	104.2%	756.5	33.6%	711.9	-5.9%
West	182.3	44.9%	447.5	145.4%	626.9	40.1%	618.8	-1.3%
Nation	1,717.9	67.4%	2,941.4	71.2%	3,347.8	13.8%	2,888.6	-13.7%

Note: Land sales income in the table is based on the actual income entering the treasury, based on Ministry of Finance statistics.

Source: Annual *Quanguo tudi churang shouzhi qingkuang* 《全国土地出让收支情况》 [Status of National Land Sales Income and Expenditure], published by the Ministry of Finance.

Table 6.2: Top 10 cities in 2012 for land sales income (units: billion yuan, %).

	2012	2011	Year-on-year growth
Shanghai	99.09	128.9	-23.10%
Wuhan	94.99	58.95	61.20%
Chongqing	89.75	102.65	-12.60%
Suzhou	72.44	87.75	-17.50%
Hangzhou	68.84	76.83	-10.40%
Beijing	64.79	105.51	-38.60%
Tianjin	64.14	82.75	-22.50%
Chengdu	63.53	43.67	45.50%
Shenyang	56.26	64.89	-13.30%
Ningpo	49.28	43.03	14.50%

Source: compiled by author from each city's statistics bureau.

As for the cities, in 2012, among the top 10 cities for land sales income, the two with the highest year-on-year growth were Wuhan and Chengdu, which were as high as 61.2% and 45.5%, respectively (see Table 6.2). Among the top 50 cities for land sales income, there were 18 cities showing positive year-on-year growth, and of these, the central and western regions each had 5, and the eastern region had 8. Among the 20 cities with the largest decline in land sales income, the central and western regions had only one each, and the eastern region had 18.

Second, the increase in land sales income in the central and western regions depends mainly on the expansion of land area sold and there is a clear tendency to consume land. From 2008 to 2012, the area of land sold in the Center increased from 41,500 hectares to 87,200 hectares, an average annual increase of 20.38%, while in the West the area of land sold increased from 37,600 hectares to 113,000 hectares, an average annual increase of 31.66%. The unit price of land sales in the West rose much less than that of the East and Center. During that period, the unit price in the Center and West increased by 3.11 million yuan per hectare and 980,000 yuan per hectare, respectively, with an average annual increase of 14.08% and 5.48%, while in the East it was 10.65%. In 2012, the unit prices for land sales in the East, Center, and West were respectively 11.827 million yuan per hectare, 7.6068 million yuan per hectare, and 5.1006 million yuan per hectare. The unit prices in the Center and West were, respectively, only 64% and 43% that of the East (see Table 6.3).

Third, industrialization in the central and western regions is driven by low-cost land. Results of monitoring by the former Ministry of Land and Resources in

Table 6.3: Area of land sold and price in the three regions (units: 10,000 ha, 10,000 yuan/ha).

	2008		2009		2011		2012	
	Land area sold	Unit price	Land area sold	Unit price	Land area sold	Unit price	Land area sold	Unit price
East	8.67	789	12.51	1,006	15.62	1184	12.27	1,182
Center	4.15	449	4.76	493	9.85	723	8.72	760
West	3.76	412	4.81	467	8.25	716	11.30	510
Nation	16.58	618	22.08	778	33.72	944	32.28	833

Sources: The annual *Quanguo tudi churang shouzhi qingkuang* (see Table 6.1).

105 key cities show that in the third quarters of each year from 2000 to 2012, the average annual increase in industrial land prices in the northwest and southwest regions was only 2.91% and 2.92%, respectively, which were lower than those in the eastern region (5.63%) and northern region (5.33%). As of the third quarter of 2012, the price of industrial land in the northwest, southwest and mid-south regions were only 586 yuan/m^2, 585yuan/m^2, and 838 yuan/m^2, respectively, far lower than the 1,086 yuan/m^2 and 931 yuan/m^2 in the eastern and northern regions, respectively.

2. Why the Land Sales Model Is Unsustainable

2.1 As the Volatility of Land Sales Income Increases, So Does Instability

From 2000 to 2012, the total contract price of land sales in China each year increased from 59.6 billion yuan to 2.69 trillion yuan, with an average annual growth rate of 37.4%, which was much higher than the average annual increases of the GDP (14.8%) and fiscal revenue (21%). However, land sales income fluctuates greatly from year to year. From 2000 to 2012, it fluctuated by more than 50% annually. Since 2010, the volatility of land sales income has clearly increased, as it has been significantly affected by the macroeconomic situation and especially by the real estate market. From 2010 to 2012, land sales income increased year-to-year by 59.87%, 14.69%, and -14.60%, respectively, and in 2011 and 2012, it decreased year-to-year by 45.18 and 29.29 percentage points, respectively (see Figure 6.1).

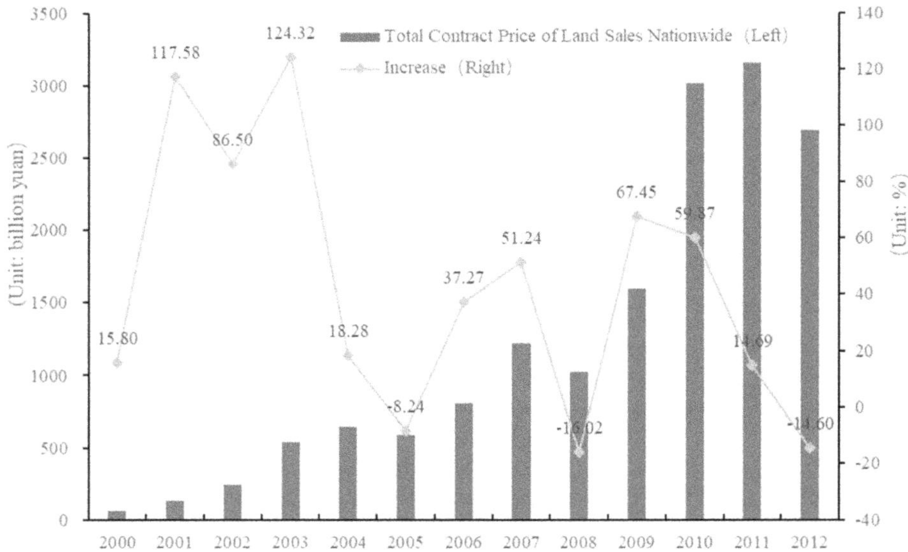

Figure 6.1: Total contract price and increase of land sales nationwide, 2000–2012. Source: *Zhongguo guotu ziyuan tongji gongbao* 《中国国土资源公报》 [*Bulletin of China Land Resources*] (annual).

2.2 As Compensatory Expenditures Rise, So Does the Cost of Urbanization

In recent years, as urbanization has accelerated, land values have risen sharply, farmers' rights awareness has been awakened, and in some areas, land use by cities and towns has gradually shifted from newly added land to land held in reserve. The cost of land expropriation and demolition has risen swiftly, and the proportion of land sales income going to compensate landless farmers and resettled residents is rising significantly. From 2008 to 2011, the proportion of total land sales income going for compensatory expenditures such as compensation for land expropriation and demolition, subsidies for land-expropriated farmers, enterprise employees' resettlement fees, and operational fees in land sales grew from 47.0% to 55.4%, with an average annual increase of about 1.4 percentage points (see Table 6.4). In 2012 especially, due to the influence of the *Housing expropriation and compensation regulations on state-owned land* (《国有土地上房屋征收与补偿条例》), the increase in compensation for land expropriation and demolition became more obvious, so that it reached 60.2% of the total land sales income.

As we may have anticipated, the government, while continuing to promote urbanization and following the provisions of the current *Law on Land*

Table 6.4: Land sales costs as a share of income (units: billion yuan, %).

	2008	Share	2010	Share	2012	Share
Land sales income	994.21	–	2,939.8	–	2,888.63	–
Land expropriation and demolition fees	366.21	36.8%	1,067.7	36.3%	1,740.16	60.2%
Subsidies for land-expropriated farmers	15.77	1.6%	44.97	1.5%		
Enterprise employees' resettlement fees	78.24	7.9%	333.66	11.3%		
Operational fees in land sales	7.11	0.7%	15.75	0.5%		
Initial land development expenditure	90.34	9.1%	256.12	8.7%	522.33	18.1%
Total Costs	557.68	56.1%	1,718.2	58.4%	2,262.49	78.3%

Note: Expenditures related to land expropriation and demolition = land expropriation and demolition fees + subsidies for land-expropriated farmers + enterprise employees' resettlement fees + operational fees in land sales. Cost expenditures = expenditures related to land expropriation and demolition + initial land development expenditure.

Sources: The annual *Quanguo tudi churang shouzhi qingkuang* (see Table 6.1).

Administration, is already having difficulty expropriating land by compensating farmers based on the original use of their land, since the compensation for the farmers' land and houses was bound to increase substantially. This situation is quite obvious in urbanized regions (including the urban-rural fringes and urban expansion zones). At present, China's urbanization is bidding farewell to the low-cost era when the compensation for land expropriation was low (see Table 6.4 and Figure 6.2).

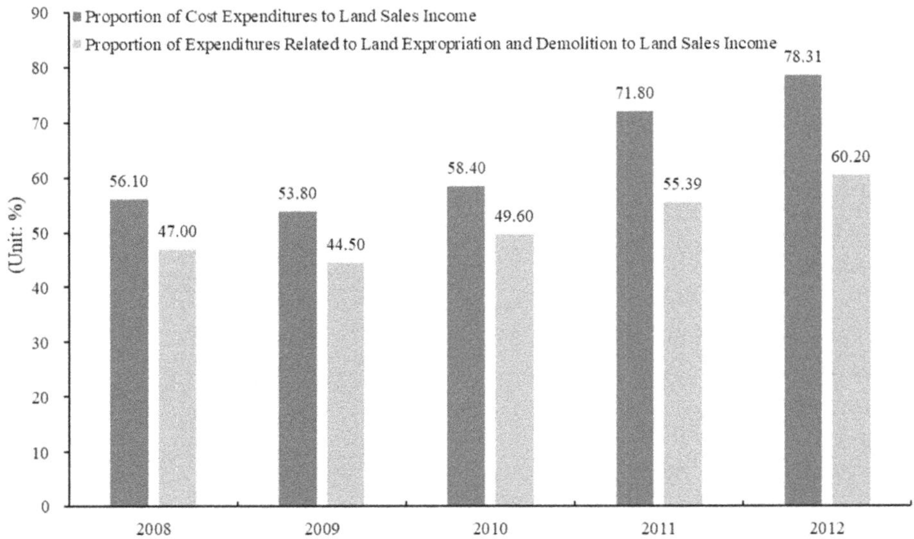

Figure 6.2: Proportion of expenditures to land sales income (unit: %).

2.3 As Local Governments' Land Revenue Decreases, Their Social Spending Faces a Challenge

In recent years, as the costs of land expropriation and demolition have increased, the actual disposable land revenue (what is called "land finance") of local governments has contracted rapidly, so that the year-on-year growth rate of that income has been declining. From 2008 to 2012, the actual disposable land sales revenue of local governments accounted for 36.32% of annual land sales income, on average. In 2008, the local governments' actual disposable land sales revenue amounted to 436.4 billion yuan, which accounted for 43.9% of the land sales income. In 2012, however, when the local governments' actual disposable land sales revenue amounted to 626.8 billion yuan, it was a year-on-year decrease of 317.2 billion yuan, and the proportion of income from that year's sales was only 21.7%. The local governments' fiscal revenue was also affected as the proportion contributed by "land finance" declined. From 2008 to 2012, the average proportion of land sales income in local governments' fiscal revenue was only 18.75%, and in 2011 and 2011, it had declined further, to 17.97% and 10.26%, respectively (see Figure 6.3).

Since 2008, China has successively introduced relevant policies to adjust the structure of land sales revenue and expenditure, tending to put more of the land sales revenue into rural construction, education, water conservancy, affordable

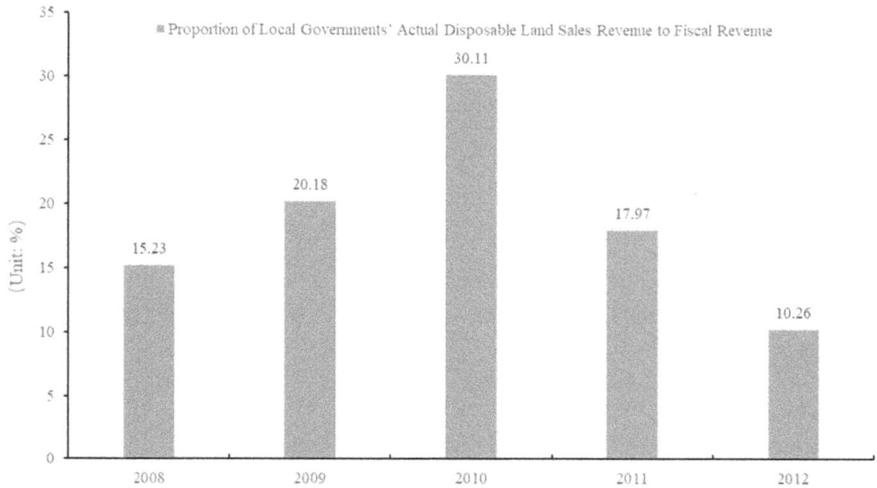

Figure 6.3: Proportion of local governments' actual disposable land sales revenue to fiscal revenue, 2008–2012.
Sources: The annual *Quanguo tudi churang shouzhi qingkuang* (see Table 6.1) and *Zhongguo tongji nianjian [China Statistical Yearbook].*

housing projects, and other areas relating to people's livelihoods. In 2012, the proportion of land sales revenue used for spending on education, farmland water conservancy construction, rural infrastructure building, and affordable housing projects were, in order, 4.66%, 3.87%, 8.42%, and 10.23% (see Table 6.5). However, this sort of institutional arrangement, which relies on "land finance" to pay for the cost of protecting people's livelihoods and other expenses, faces the problem of sustainability. First, land sales revenue is quite volatile, but social spending is rigid. Once land sales revenue falls sharply, local governments will have difficulty fulfilling their commitments on social spending. Second, the various types of cost expenditures in the land expropriation process have risen rapidly, and in the future it will result in a continuing decline in the proportion of land sales revenue to total income. This will directly affect the ability of land sales revenue to defray the cost of protecting people's livelihoods.

2.4 Shrinking Land Sales Income Means Less Funding for Urban Construction

In recent years, China has made great achievement in urban infrastructure building, and land sales income has been an important funding source for it. Between

Table 6.5: Amount and proportion of land sales revenue used for various expenditures (units: billion yuan, %).

	2008	Proportion	2010	Proportion	2012	Proportion
Land sales revenue	456.308	–	1,221.599	–	626.141	–
Education	–	–	–	–	26.995	4.66%
Farmland water conservancy construction	–	–	–	–	22.459	3.87%
State-owned Land Revenue Fund	30.891	6.74%	100.789	9.00%	101.717	17.55%
Agricultural land development	12.607	2.75%	18.887	1.69%		
Land use fees for additional land designated for construction projects	63.871	13.94%	98.373	8.79%		
Affordable housing projects	14.557	3.18%	46.362	4.14%	59.301	10.23%
Urban construction	302.402	66.0%	753.167	67.26%	320.415	55.27%
Rural infrastructure construction	33.83	7.38%	102.168	9.12%	48.808	8.42%
Land sales revenue used for expenditures subtotal	458.158	–	1119.746	–	579.695	–
Land sales revenue balance	-1.85	–	101.853	–	46.446	–

Notes: (1) Land sales revenue is the balance of the current year's land sales income after deducting cost expenditures (see Table 6.4).

(2) Expenditure on education, farmland water conservancy construction, the State-owned Land Revenue Fund, and agricultural land development are all special funds, with the balance carried over for the next year's use, according to regulations. Therefore, the listed figures are that year's actual accruals.

(3) Land sales revenue balance is the difference between the land sales revenue and listed categories of expenditure from the revenue.

(4) Expenditure for education and farmland water conservancy construction was money accrued from 2011.

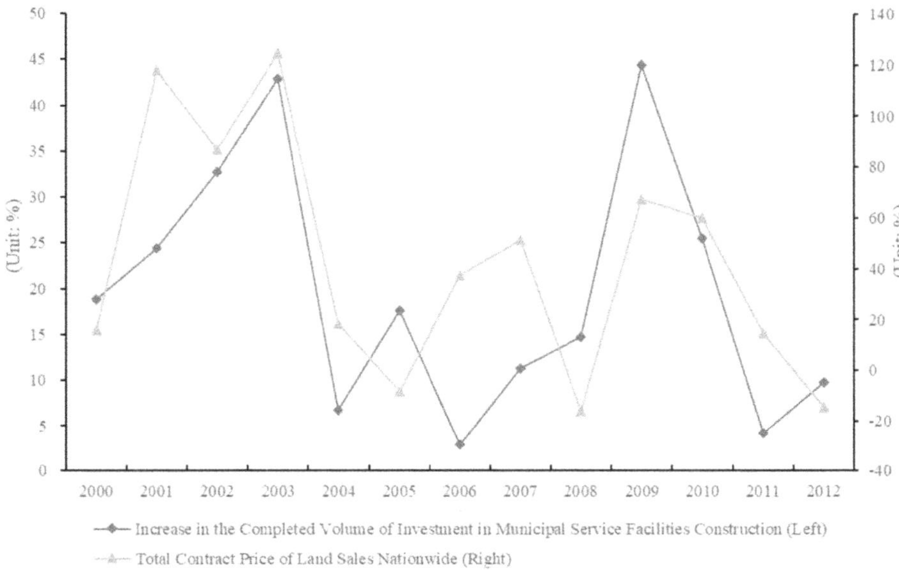

Figure 6.4: Relationship between increase in land sales income and increase in the completed volume of investment in urban service facilities construction, 2000–2012. Sources: The annual Zhongguo guotu ziyuan gongbao [China National Land Resources Bulletin] and Zhongguo chengshi jianshe tongji nianjian [China Urban Construction Statistical yearbook].

2000 and 2012, there was a significant positive correlation between the increase in land sales income and the increase in the completed volume of investment in urban service facilities construction, and the fluctuations were relatively close (see Figure 6.4). We can view these in stages. From 2000 to 2003, the annual growth rate of completed volume of investment in urban service facilities construction continued to increase, with an average annual growth rate of nearly 30%. The average annual growth rate of land sales income during that period was 86%, the highest increase since 1998. From 2008 to 2012, the year-on-year growth rate of land sales income showed an inverted U-shape of "first rise, then fall." In 2009, land sales income increased by 67.45% from the previous year, then it continued to drop until 2012, when it hit -14.6%. During the same period, the year-on-year increase in the completed volume of investment in urban service facilities was entirely in line with the growth trend in land sales income. In 2009, the completed volume of investment grew 44.42% from the previous year, which was the highest point in recent years, but since 2010, the year-on-year growth rate has gradually declined (rose in 2012) (see Figure 6.4).

With the slowdown in land sales income growth and the reduction in land sales revenue from rising cost expenditures, the urban infrastructure-building model that relies heavily on land sales revenue will face severe challenges in the future. In the current land income/revenue and expenditure structure, there are two main funding sources for urban construction: the first is expenditure from land sales income under the heading of initial land development, and the other is expenditure from land sales revenue under the heading of urban construction. From 2008 to 2012, an average of 12.3% of land sales income went to initial land development, and more than 55% of land sales revenue each year went to urban construction. In recent years, the growth trends of the two proportions have differed. On the one hand, the proportion of initial land development expenditure to land sales income has been rising steadily. In 2012, that proportion was 18.1% (compared with 9.1% in 2008). On the other hand, the proportion of urban construction expenditure to land sales revenue is declining. In 2012, 320.415 billion yuan from land sales revenue was spent for urban construction—a decrease of 236 billion yuan from the previous year. It accounted for 55.3% of the actual land sales revenue that year, down 10.7 percentage points from the previous year. In the future, if this model of relying on land sales to promote urban construction is not changed, urban construction and maintenance will face financial sustainability problems.

2.5 As Land Mortgage Loans and Illegal Mortgage Financing Increase, So Does Risk to Local Governments

In recent years, both the land area in China that has been used as collateral for loans and the volume of the loans have continued to rise rapidly, and the volume of these loans as a share of RMB loans from financial institutions nationwide has also gradually increased. From 2007 to 2012, the area of land collateralized in 84 key cities increased from 128,400 hectares to 349,000 hectares, with an average annual increase of 44,100 hectares. The volume of land mortgage loans taken out increased from 1.33 trillion yuan to 5.95 trillion yuan, with an average annual increase of 34.8%. From January 2008 to November 2012, the proportion of land mortgage loans to RMB loans of financial institutions nationwide increased from 4.94% to 9.29%. In 2012, various ministries and commissions successively issued several policies to prevent financing platform risks. As a result, the collateralized land area and loan growth rate that year decreased to 15.7% and 23.2%, respectively, from the previous year, which was the lowest growth rate since 2007 (see Figure 6.5).

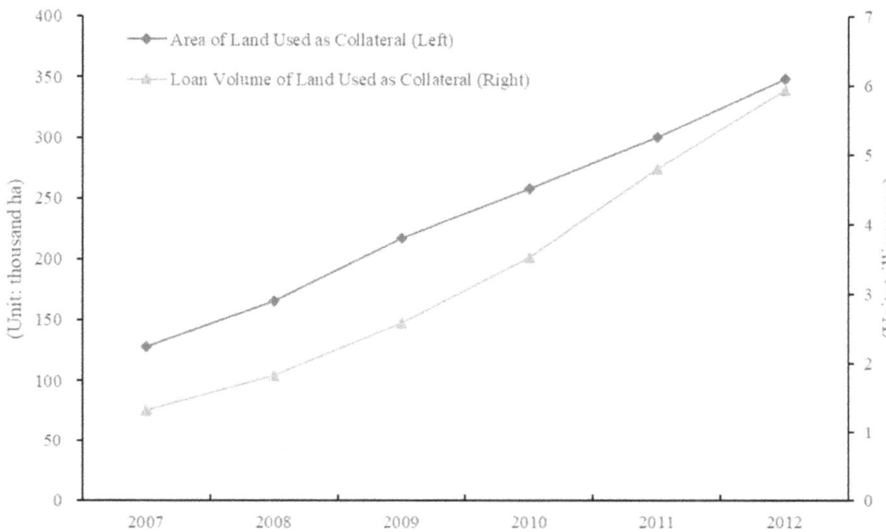

Figure 6.5: Area of land used as collateral and loan volume in 84 key cities, 2007–2012. Sources: The annual *Zhongguo guotu ziyuan tongji gongbao* [China National Land Resources Bulletin].

Let us examine the regional differences. In 2012, most of the 5.95 trillion yuan that the 84 key cities borrowed, using land as collateral, were concentrated in 30 cities in the East (4.02 trillion) and 14 cities in the West (1.08 trillion), followed by 21 cities in the Center (610 billion), and finally 19 cities in the Northeast (240 billion). These accounted for 67.6%, 18.15%, 10.25%, and 4%, respectively (Figure 6.6). As for the cities themselves, the three cities with the largest land mortgage loans were Beijing (1.4194 trillion), Shanghai (562.4 billion), and Chongqing (478.3 billion); accounting for 23.8%, 9.4%, and 8%, respectively, of the land mortgage loans to the 84 key cities. As of November 2012, the proportion of land mortgage loans in Beijing and Chongqing accounted for 38.9% and 31.8% of their local RMB loan balances, respectively.

The risk on land mortgage loans first appears as local government debt risk. According to National Audit Office reports, at the end of 2010, local governments had pledged to use land sales income to repay their debt balance of 2.5 trillion yuan, which is 81.2% of the outstanding debt balance for which they are responsible for repaying. If we look at debt repayment year, we find that the debts repaid in the three years from 2013 to 2015 accounted for 11.37%, 9.28%, and 7.48%, respectively, and 30.21% were due during and after 2016. Due to the increased instability of land sales income in recent years and expected future

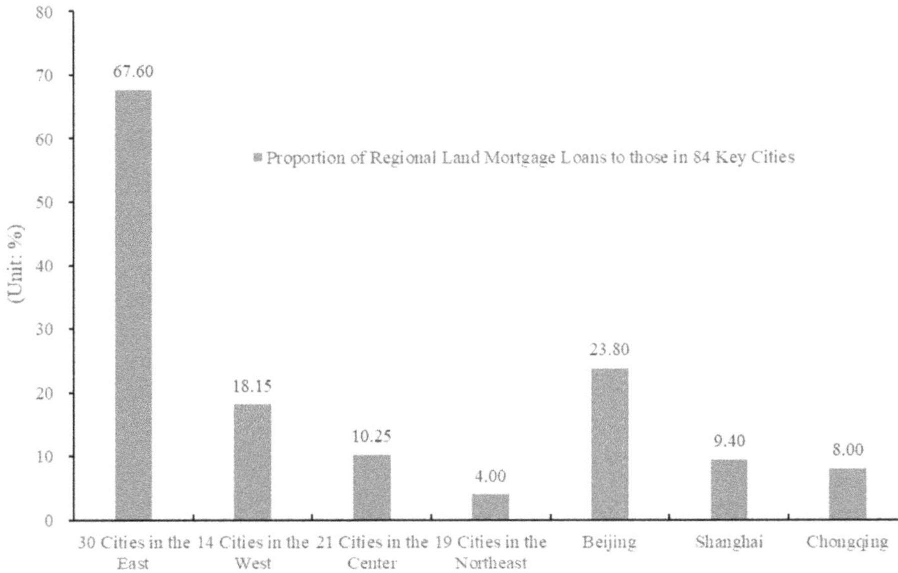

Figure 6.6: Regional analysis of land mortgage loans in 84 key cities in 2012.
Sources: The annual *Zhongguo guotu ziyuan gongbao* [Bulletin of China Land and Resources].

volatility, and the declining actual land-based disposable revenue that local governments are receiving, the potential debt risk of local governments (especially county and city governments) will significantly increase if they use land sales income as the major source of repayment.

In comparison with local governments' debt risk, though, what needs to draw more attention is the financial risk of land mortgage loans. At the end of 2010, the government debt balance of the financing platform companies was 4.97 trillion yuan, which accounted for 46.38% of the balance of local government debt. The most important collateral that the financing platform companies use is the various kinds of reserved land. If someday the land market were to have a large fluctuation, then, by affecting the value of the collateralized land, the financial risks of the various banks may accumulate rapidly.

A number of classic surveys have shown that in land-collateralized financing there have been extremely opaque financing processes and a large number of illegal operations, for example, changing the type of land use rights and ownership registration, putting a false use registration on land whose value would otherwise be difficult to realize (such as public facilities land or public interest land), illegally reserving land or acting as third-party guarantor, registering land and using it as collateral even though the legal use right for it has not been obtained, duplicate registrations, using the same land twice as collateral for loan financing, using falsely high valuations to increase the volume of loan financing, and so on. In 2012, a certain agency responsible for monitoring state-owned land investigated the land financing situation in six cities and found that approximately 2/3 of the loans, in terms of number of loans, the area of land involved, and the volume of the loans, involved some illegality. If this situation is not regulated as soon as possible, the potential risk from these land mortgage loans is very large.

3. Response Policies

In the next strategic opportunity period, China will bear the historical task of carrying out industrialization, urbanization, and agricultural modernization simultaneously. To this end, it will be necessary to build a new relationship between land and the economy, promote a transformation in land use through reform of the land system, optimize land use structure, promote land use efficiency, protect farmers' land rights and interests, balance the interests of all parties, and provide a reasonable amount of land and financing for sustainable urbanization.

3.1 Optimizing Land Structure, with Reform of the Land System as the Core

First, in view of the excessive amount of industrial land in China, we should formulate planning adjustments to convert that land from industrial use, handle the ownership and interest distribution issues, and innovatively switch industrial land to the service industry, residential land, or high-end manufacturing, to optimize the structure of urban land use, restructure industry, and achieve sustainable urban development.

Second, considering that the government's land use ratio is too high, and that its land is not intensively used, rules for allocating land by government and administrative units to the market and methods for disposing value-added income from the land should be formulated. Land for industrial development and livable space for urban use should be increased. Gains from increased land values, brought about by use conversion, should be better used for urban development.

Third, considering the extralegal, low-end uses to which farmers' land in the urban-rural fringes are being put currently, a pilot project to integrate collective construction land into the market should be carried out on a selection of such areas from different types of cities. This would release a huge amount of land for urban development and industrial upgrading.

Fourth, pilot projects should be carried out in a selection of different types of areas wherein a mechanism is set up to grant permanent urban residency to the migrant population and guarantee paid exits of rural households' homesteads. This would bring about a win-win situation in terms of rural household's homestead usufruct and the optimization of urban-rural land use.

Fifth, considering that in the eastern coastal areas, future industrial upgrading and urbanization will switch to using mostly land held in stock, it is recommended to focus on carrying out pilot programs and innovations in systemic reform to optimize the land structure in these areas.

3.2 Simultaneous Reform of the Land Expropriation System and Release of Collective Construction Land into the Market

In the amendment draft of the *Law on Land Administration*, which focuses on the reform of the land expropriation system, a relatively mature and feasible plan has already been formed. It changes "original use compensation" to "fair compensation," which would not only bring China's land expropriation system closer to the principle of land expropriation as found under market economy conditions but also reflect the locality and the asset-quality of the expropriated land and realize the right to develop it; it will be conducive to expropriating land in accordance with law. In the resettlement system, the housing resettlement of land-expropriated farmers should be separated from compensation and resettlement, and market-based compensation should be implemented, the local experience should be summed up and the social security resettlement should be increased. This will be more conducive for ensuring the long-term livelihood of landless farmers.

Considering that misinformation on the various revisions to this amendment now circulating in society is seriously affecting the normal implementation of land expropriation in the process of local industrialization and urbanization, it is recommended that the National People's Congress should pass and implement the existing draft as soon as possible. At the same time, in order to ensure the rational use of land for local industrialization and urbanization, it is recommended that the pilot project experiences of Guangdong and other places be summarized and that farmers' collective construction land, under planning and use controls, be allowed to enter the market. Since the central and western regions are still in the rapid industrialization stage, the entry of rural collective construction land into the market is especially urgent.

3.3 Promoting Land Asset Management and Financial System Reform, to Provide Sustainable Funding for Urbanization

First, modes of state-owned land asset management should be explored. The government's land administration function should be separate from profit-oriented land management, and the profit-oriented management function of state-owned land should be separated from the work of current government agencies. On the basis of clarified status quo and details on state-owned land, land asset management companies, managing state-owned land, should be established, and they should be engaged in managing, preserving the value of, and increasing the value of state-owned land. It should be clearly specified that the business income of such companies enters the local treasury, and that the use of the funds should be decided and monitored by the local People's Congress.

Second, the state-owned land financing system should be refined. Certificates on state-owned land rights should be strictly managed, categories of loans using state-owned land as collateral should be clarified, the independence of third-party assessments of land value should be enhanced, punitive measures for violations of regulations on land mortgage loans should be instituted, and the security of loans with state-owned land as collateral should be guaranteed.

Third, with reference to the *Regulations on Implementation of Zone Expropriation* (《区段征收实施办法》) in the Taiwan region, the land use model should be reformed so as to reduce the dependency of various levels of government on land sale mechanisms and on financing through financial platforms.

Fourth, the pilot land property tax system should be differentiated. For reserved land, land held in store by real estate businesses, and rural collective construction land entering the market, collection of the pilot land value-added tax should be implemented first.

7

Thoughts on China's Land System Reform and How to Promote It

1. The Characteristics and Problems of China's Land System

In comparison with other countries and regions around the world, China's land system is unique.

1.1 The Land Rights System in Which Ownership and Use Rights Are Separated on the Basis of Socialist Public Ownership

China's land rights system is built upon four elements: First, the implementation of socialist public ownership of land. No organization or individual may misappropriate, buy, sell, or otherwise illegally transfer land. Second, the system of state ownership of land and the system of collective ownership of land coexist. Urban land and natural resources such as mineral deposits, water currents, sea areas, forests, mountains, grasslands, wastelands, and tidal flats are owned by the state. Rural land and land in suburban areas of cities, homesteads, self-reserved land, and self-reserved mountains, except where laws and regulations stipulate state ownership, are owned by collectives; where laws and regulations stipulate collective ownership, land, forests, mountains, grasslands, wastelands, and tidal flats, are owned by collectives. Third, landownership is separate from land use rights. On rural land, a contract management system has been adopted, and it gives farmers long-term and guaranteed land use rights. As for state-owned land, the land-use rights holders have the right to legally occupy, use, derive income from, transfer, lease out, and use as collateral their land use rights, under the premise of state ownership. Fourth, land property rights are respected and protected. The state regulates and protects land rights in accordance with the *Constitution*, *Real Right law*, *Law on Land Administration*, and *Rural Land Contract Law*. Provided that the ownership of the land does not change, the land rights of land users are protected, with focus on improving and strengthening usufruct of land.

1.2 Land Administration That Aims to Protect Arable Land and Is Centered on Use Regulation

Because of its high population density, China gives first place to food security and the protection of farmland. When the *Law on Land Administration* was amended in 1998, the effective use regulation systems of foreign countries were introduced and written into the law, and it was proposed that it was proposed that "to cherish and rationally use land and effectively protect cultivated land are the basic state policies of China," The most stringent farmland protection system was implemented through a series of institutional arrangements. Through the preparation of master planning of land use, the delineation of land use areas, the determination of land use restrictions, and the conversion of agricultural land, especially cultivated land for construction, shall be strictly restricted.

1.3 A Government-Guided Land Resource Allocation Method That Mainly Follows Market Mechanisms

The content of China's land resource allocation system includes the following: First, for state-owned land, a dual-track system has been instituted that supplies land via planning or by paid use. Apart from land used by government agencies, the military, urban infrastructure, public utilities, and major state-supported projects for energy, transportation, hydrology, and other infrastructure, for which land is supplied via planning, all other types of construction land are supplied via different forms of paid use, that is, sale, leasing, price contribution, or shareholding. Second, land used for profit-oriented operation is subject to market allocation under the government's exclusive monopoly. For profit-oriented land—that is, industrial, commercial, tourism, entertainment, commercial residential land, and so on—and for land parcels having two or more parties that are willing to use them, the government sell the use right of land by means of Land bid invitation, auction and listing, open competitive bidding, etc. Third, land use rights holders on state-owned land can lease out the land, arrange for shareholding in it, transfer it, or use it as collateral.

1.4 A Land Administration System Based on Centralized and Unified Management

Before 1986, the administration of land in China was decentralized, managed by many departments. In 1986, as a response to problems arising from the "block-based" land administration system, a unified management system for urban and rural land administration was established. In 1998, the system whereby different levels of the government had quotas within which they had authority to review, approve, and administer (including expropriate) land was replaced by land use regulation. That is, the authority to review and approve conversion of the use of agricultural land and to expropriate land was limited to the State Council and provincial-level governments. In 2004, the state-owned land administration system below the provincial-level government was further put into order. In 2006, the National Land Supervision System (国家土地督察制度) was established. By this point, the unified land administration model for the relatively centralized management and top-down supervision of land had basically taken shape. That is, Administrative department of land and resources under the State Council is responsible for the overall management and supervision of land throughout the country, and it implements four levels of management (central, provincial, municipal, county). Under a team leadership management system, it practices a double level of management (central and provincial), with vertical management below the provincial level.

With economic and social development, however, this land system showed the following major drawbacks:

(1) In the property rights arrangement, there exists a duality of rights and an impairment of powers. Under the socialist public ownership system, collective ownership of rural land and state ownership of urban land was established, and within that system, rural collective members only have the right to use, revenue, and transfer the contracted land for agricultural purposes. When the agricultural land was converted into urban land, the expropriation, transfer, and assignment of that land was handled

by the municipal and county governments, and the various land-using units (local governments, business enterprises, units, etc.) had the right of management, usufruct, and transfer under the control of the planning once they had obtained the right to use land. Thus the two types of landownership entailed two completely different kinds of land rights, which were governed by different legal systems, and the government in particular became the owners and property controller of cities.

(2) There is inequality of access to the land market for different entities. Since the land market has an urban-rural division, and rural collective land is bordered by village communities, collective members may access it. Agricultural land circulates mainly through informal transactions without any price, and different types of agricultural land (contracted land, homesteads, collective construction land, non-cultivated land) enter the market according to different rules of access. The supply, transfer, and recovery of urban land, meanwhile, are monopolized by the local government. Land transactions are subject to competition among buyers, under the seller's monopoly, and different types of land (public purpose land, industrial land, profit-oriented land) are assigned in different ways. This results in land price distortions and inefficient resource allocation.

(3) Gains and losses in land value-added income are unfairly distributed among relevant stakeholders. The principle of land value-added income distribution is such that land-expropriated farmers receive compensation based on multiples of income from the land's agricultural use; the land value-added income is obtained, used, and controlled by local governments when the land use is converted; and future land value-added income returns mainly to the land occupiers and partly to the local governments. As a consequence, the compensation legally received by the original collective owners is too low, and the compensation received by urbanized farmers is non-standardized, opaque, and unprincipled. Meanwhile, the one-time land value-added income that the government gets from the land is too high, but the future land value-added income is lost. The one-time land assignment fees paid by real estate businesses and homebuyers are too high, but the future land value-added income that they will derive from it is also too high. This is one of the institutional factors that contribute to the current unfair distribution of income.

(4) There are conflicting goals and functions within the land administration system. Despite the use controls and planning controls in the land administration system, under the specific implementation measures that

> apply, land use is controlled through a master plan, annual plan quotas, and approval management. However, due to defects in the planning system and excessively strong government administrative leadership, there is a conflict between the goals of protecting arable land and promoting development, and there is gamesmanship between the central and local levels. Use regulations give way to planning regulations, and planning regulations plus the land expropriation and conversion system lead to ownership controls. The system of review and approval and annual planning management shows inefficiency, disconnection from development needs, and rent-seeking, corrupt behavior in connection with land.

A common type of land use control is *the planning unit development system*, which is a zoning system closely related to average density; it is also a zoning system for comprehensive development or clustering. Developers can mix different types of building uses and building densities, as long as they follow certain regulations on overall density and land use.

Box 1: The Main Features of Land Use Zoning Regulation in the West

The basic idea behind land use zoning control is to "divide land into areas of different use, stipulate different land regulation rules for different areas, and implement the identical regulation rules within each area." The zoning regulation system for land use was first established in 1692, and since the 1940s and 1950s, it has become an increasingly popular means of planning and managing land in most countries and regions. The main purposes for implementing zoning controls on land include the following:

(1) **Excluding improper land use and avoiding "external diseconomic" land use.** Zoning controls on land use eliminates misallocation of land resources or "external diseconomic" effects caused by private tenure of land. Furthermore, it protects and improves people's living environments by separating them from inappropriate forms of land use; it also protects existing buildings, the natural heritage, and rural landscapes, thereby improving the environmental standards of where people live and ensuring the realization of health and safety goals.

(2) **Controlling land use and speed of development effectively and systematically, and ensuring the maintenance of public facilities and services.** Zoning regulation of land not only regulates location and type of land use and development activity but also limits the scale, intensity, and population density of land use, so that overall land use and economic development are effectively controlled, and a certain service level of public facilities can be effectively guaranteed.

(3) **Economizing on government spending.** Zoning on land use can control land development by designating development areas and guide necessary construction of housing, factories, service facilities, schools, and so on into the appropriate places, thus ensuring that land may be used to full effect, the cost of public services may be reduced, and the government may economize on its expenditures.

2. China's Current Land System and the Transformation of Development Methods

During the most recent golden opportunity for economic development, China used its unique land system to ensure rapid economic growth, industrialization, and urbanization. However, this system also shaped a land-supported economic development model that was uncoordinated, unhealthy, and unsustainable.

(1) Land became a tool for sustaining traditional forms of development. China's economic growth depends on an ample land supply. From 2003 to 2012, the total supply of state-owned construction land increased from 286,400 hectares to 690,400 hectares, with an average annual growth rate of 10.27%, and the GDP growth rate was highly correlated with the land supply. High-speed industrialization relied on the maintenance of industrial land at low cost. From 2000 to 2012, when the average annual increase in the national comprehensive land price level, commercial land price level, and residential land price level was 10.04%, 11.40%, and 14.42%, respectively, the average annual price increase for industrial land was far less—only 3.35%. Rapid urbanization was driven by land expansion and land capitalization. Between 2000 and 2012, the built-up area of cities more than doubled, and land sales income increased by more than 45 times. In 2012, the balance of loans in 84 key cities that had used land as collateral reached 5.95 trillion yuan, so land became an important source of funding for urban infrastructure building.

(2) "Development driven by land" has latent economic and social risks. First, the repayment of debt by local governments is rather highly dependent on land sales income. By the end of 2010, of the debt balance of local governments with repayment obligations, the balance of debts for which the local governments had pledged to use land sales income as the source of their repayment was 2.5 trillion yuan, which accounted for 81.2% of their debt balance. In terms of debt service year, the debts they repaid during the three years from 2013 to 2015 accounted for 11.37%, 9.28%, and 7.48%, respectively, and the remaining 30.21% was repaid after 2016. Due to the increased instability of land sales income in recent years, the proportion of land income that the government actually has at its disposal is still declining. Therefore, the use of land sales income—the uncertainty of which will increase in the future, and which is generating

less and less disposable income—as the most important funding source for debt repayment, will clearly increase the potential debt risk of local governments, especially county and municipal governments. The second is bank financial risk. Banks are the largest providers of funds for local government debt. At the end of 2010, bank loans accounted for 79.01% of local government debt funding, and of this amount, 74.84% were bank loans with debt repayment obligations. The government debt balance of financing platform companies accounted for 46.38% of local government debt balance. The main way for local governments to obtain loans has been to use land as collateral. From 2007 to 2012, the area of land in 84 key cities that had been used as collateral for loans increased from 128,400 hectares to 340,900 hectares, with an average annual increase of 44,100 hectares. The amount of loans taken out with land as collateral increased from 1.33 trillion yuan to 5.95 trillion yuan, with an average annual increase of 34.8%. From January 2008 to November 2012, land mortgage c as a share of RMB loans from national financial institutions increased from 4.94% to 9.29%. The third is social risk. Among group petitions, 60% relate to land, and 40% of all community-related petitions are land disputes. Of these, 84.7% of land disputes relate to compensation for land expropriation. Each year, there are as many as around 4 million disputes caused by land expropriation and housing demolition.

(3) Compensatory expenditures for land expropriation, house demolition, and so on have increased substantially, and urbanization costs have risen. In recent years, land values have risen sharply, farmers have become more aware of their rights, and in some areas urban land use is shifting gradually from newly-added urban land to land held in store. The cost of land expropriation and demolition has increased rapidly, and the proportion of land sales income that goes to compensate land-expropriated farmers and those whose houses have been demolished are increasing rapidly. From 2008 to 2011, compensation for land expropriation and demolition, subsidies for land-expropriated farmers, enterprises employees resettlement fees, and operational fees in land sales rose from 47.0% of total land sales income to 55.39%, with an average annual increase of approximately 1.4 percentage points. In 2012, the expenses of land expropriation and housing demolition even rose as high as 60.2% of total land sales income. The government follows the provisions of the current *Law on Land Administration* (and is expected to do so in the future),

but it has already become difficult to expropriate land while offering compensation based on the land's original use. China's economic development is bidding farewell to the era of low-cost compensation for land expropriation.

(4) Scarce land resources are not being used intensively and economically. First, with regard to urban expansion, from 2000 to 2011, the total urban built-up area in the country increased from 22,439 square kilometers to 43,603 square kilometers, with an average annual growth rate of 6.225%. From 2000 to 2010, the average urban population growth rate was 35.31%, while the growth rate of the urban built-up area over the same period was 99.29%. In other words, the urban built-up area was growing 3.95 times faster than the urban population. Second, the land is being used too freely and extensively. For many years, more than 40% of the construction land nationwide has been used for industry, mining, and warehousing, thus "crowding out" residential and living space, but the plot ratio of industrial projects has ranged between only 0.3 and 0.6. Third, the nonmarket allocation of urban land leads to the creation of image projects, such as large squares, wide roads, and large office buildings. Fourth, rural-urban fringe in different kinds of cities occupy a great deal of land, but because they exist in a situation of extrajudiciality, they are put only to inefficient use. Fifth, rural construction land occupies too much space, and the area of land designated for rural residents' use is as much as 170,000 square kilometers.

(5) Much high-quality arable land is being occupied, which endangers national food security. The accelerated pace of the last round of industrialization and urbanization in coastal areas led to the loss of a lot of cultivated land. From 1996 to 2008, a total of 2,703,700 hectares of cultivated land was occupied by construction, and the net area of cultivated land decreased by 831.71 million hectares. The traditional pattern of "grain from the south is transported north" has changed to "grain from the north is transported south." Since the mid-1990s, China's grain supply has basically been dependent on the north. Of China's 31 provinces and municipalities, half of the 13 major grain-producing areas have been unable to allocate food out, and five of the seven main sales areas are in the south. As industrialization accelerates in the central and western regions, the trend of occupying the high-quality cultivated land continues to spread. If the land use pattern of industrialization and

urbanization is not changed, the supply and demand pattern for food in China will further reverse, and the threat to national food security cannot be ignored.

Clearly, China's traditional development model is highly dependent on land. The high-input, high-consumption factor input methods are mainly seen in the ample supply and high level of consumption of land. Competition among local governments depends mainly on their use of land to attract business investment. Rapid industrialization depends on the supply of low-cost land, at distorted prices; rapid urbanization depends on the capitalization of land under the government's exclusive monopoly. To change the traditional methods of economic development, we must change the way land is used.

3. The General Idea and Main Content of Reform

(1) *Direction of reform*. Reform of the land system will determine the success or failure of the transformation of the methods of land use and economic development. It will also be the key to whether we can make good use of the next round of development opportunities and promote national strength to a higher level. The fundamental goal of the land system reform would be to provide institutional guarantees for transformed economic development and sustainable urbanization. The basic direction would be to establish the following: (a) a land property rights system wherein the rights appertaining to urban state-owned land and rural collective land are equal, in line with the spirit of the equal property rights protections in joint development under many kinds of ownership systems under the *Constitution*; (b) a land market with equal access and fair trade in both urban and rural land; (c) a fair, shared land value-added income distribution system; (d) a land property tax system that befits modern social development; (e) open, transparent, and standardized systems for state-owned land asset management and for land financing; (f) a modern land administration system centered on ownership management and planning and use control.

(2) *The general idea*. The general idea would be to consider reform of the land system, that is, eliminating the dual system of urban versus rural land, as a breakthrough, then: (a) promote the optimal allocation and flow of production factors in urban and rural areas, so as to realize the integration

of urban and rural development and realize sustainable urbanization; (b) build a property rights system wherein the rights appertaining to ownership of the two kinds of land are equal, so as to achieve equal participation and sharing of economic development opportunities for different ownership entities; (c) implement confirmation and registration of real property rights, so as to provide basic institutional guarantees for safeguarding rights and interests on land, for agricultural modernization, and for urbanization of the population; (d) promote reform of the land expropriation system and of the land use means, promote the fairer distribution of land income, and realize the sharing of land value-added income by all of society; (e) bring collective construction land into the market and expand the marketized allocation of land in urbanized areas, so as to realize equal market access by different ownership entities and make more economic, intensive use of scarce land resources; and (f) improve profit-oriented land management, financing and taxation systems so as to provide financial support for sustainable urbanization.

(3) *The main content of reform.* The main drawbacks of the currant land system are the duality of rights, market access inequality, price distortions, and unfair distribution of value-added income. The fundamental goals of the land system reform will focus on equal rights, open access, and fair sharing. A land system with equal rights appertaining to the two types of landownership, a unified market, and fair sharing of value-added income should be established, through which the transformation of land use patterns and economic development patterns are to be promoted.

First, improving the rural land property rights system. A long-term, unchanging system for land contracting relationships should be implemented, with the starting point of collective land contracting clearly specified, and a system for the long-term contracting of land use to farmers should be clearly defined. We should improve policies and relevant laws in a way to establish a system in which land contracting rights and land management rights can be separated, and in which land contracting rights and management rights are equally protected. The land transfer system and the methods used should be improved, so that the principles of voluntary action, legality, and compensation are followed; land rents should return to the original farmers holding the land-contracting rights, and in any land transfer, a contract must be signed with the original contracting farmer. For improved land use rights, establish the right of disposal of land contract, and contract right and management right can be used as collateral.

Box 2: "Over a Long Term without Change" in Wayao Village (瓦窑村)

After the Third Plenary Session of the 17th Central Committee (October 2008), Wayao Village (瓦窑村) in Shuangliu District, Chengdu City (成都市双流区), Sichuan Province, took the lead in implementing the "over a long term without change" clause in rural land contractual management rights, thus creating a "national precedent." The specific practices of Wayao Village were to organize and set up a village council and form relevant system regulations. These included:

(1) **Solidifying membership of the collective economic organization.** This occurred in three steps. First, the members of the collective economic organization used the community (*she* 社) as a unit, and the households that confirmed their rights in the community in 2008 were taken as the standard by which those households, as units, enjoyed rural land contractual management rights in their community/rural village. Second, new persons in the community unit who confirmed their rights in the community during a one-year solidification period from 2008 to 2009 were recognized as special members of the collective economic organization, and they could only enjoy a distribution of collective asset income. Third, the solidification period was separately determined, with the community as unit. New persons who arrived in the community after the solidification period were no longer recognized as members of the collective economic organization.

(2) **Implementation of rural land contractual management rights "over a long term without change" (*chang jiu bu bian* 长久不变).** First, based on the starting time of the long term, unchanging land contractual management rights as determined by the various communities, no further adjustment to land contractual management rights was to be carried out. Farmers had the right to mortgage, transfer, and handle other such aspects of land contractual management rights. Second, if the contracted

land was damaged by a natural disaster (floods, mudslides, etc.), the community collective would no longer make an adjustment and allocate land to the affected contractor. Third, if a farmer had built a new house on contracted land before the confirmation of rights, then, after his original house site was restored to cultivation, an area equivalent to the new house would be contracted to the farmer's household; any excess area would then become collective construction land, and any shortfall would be deducted from the household's contracted land area.

(3) **Clarifying the way land income was distributed.** First, income from land such as the raised paths between rice paddies, riverbanks, barren slope land, and so on, that had not been distributed to any household, would revert to collective ownership and be distributed by community unit shares to the people. Such shares could be inherited, transferred, donated, etc. Second, when the contracted land of a community member needed to be used for collective public welfare undertakings, the method of deciding "who benefits and who pays" was adopted, and then the beneficiary households would have to compensate the "losing" households according to the national standards for compensation in land expropriation. The third concerns the handling of land contractual management rights and land income rights after the death of a farmer of a "five-guarantee household" (*wubaohu* 五保户). After the consent of the village committee is obtained and the matter is reported to the relevant departments of the town and county-level governments, the method of deciding "who will be supported, and who will benefit" was adopted. Fourth, average contracted area per capita in each community was calculated. Any excess in the difference between the actual measured area per household and the area of the second round contracted accounts would be cultivated by the original contracting household, and any shortfall would be made up by way of agreement by the original contractor. The end date of cultivation would be determined either by the date the land was expropriated by the state or the date the project was transferred of the land. Income from cultivation would return to the farmer.

> Any income produced at the time the land is expropriated by the state or at the time the project was transferred of the land would return to the persons holding the contracting management rights at the time when their rights had been confirmed. Various subsidies, like money from the Farmland Protection Fund (*geng bao ji jin*耕保基金) and state subsidies, that are paid to farmers based on a calculation of their land area, would be enjoyed by the contractual management rights holders as determined at the time of confirmation.

Second, providing the institutional infrastructure to protect land property rights. China's current land property rights guarantees have major flaws. The ownership entities in land collectives are unclear, powers are not clearly demarcated, and the rights protections are not strong enough. The registration and ownership management of land, housing and natural resources are still handled in separate departments, and since the registration authorities are not uniform, the certificates do not match one another, and this is not conducive to the protection of land rights, nor to the management of land. The central government must resolve to complete the identification, registration, and certification of all land resources in rural areas, ensure that confirmation of land rights to rural land resources reaches the hands of each farmer, and that basic institutional services are provided in readiness for the next stage of agricultural transformation and population urbanization. Laws and regulations on such unified registration of real estate should be promulgated as soon as possible, in line with the requirements of the *Real Right law*. Then the unified land-based registration system for real estate can be implemented, with uniform legal basis, registration authority, and ownership certificates.

Box 3: Land Registration Systems around the World

Basically, there are three different types of land registration system in the various countries and regions around the world: the contract registration system, rights registration system, and Torrence registration system.

(1) **The Deed registration system.** Under this system, as long as the two parties to a change of land rights have expressed agreement and have concluded a contract, then the legal consequences of the change of land rights occurs. The contract is the essential, effective element. The land registration is only an essential element in confrontation with third parties, that is, unregistered land rights cannot be used in opposition to any third party. The land registration authority does not conduct substantive examination of the land registration application—only a formal one, in full accordance with the documents submitted by the land registration applicant—do not ask whether the land rights are real or not. Whether the change of land rights is registered or not, it occurs in line with the intentions of the two parties, and the government places no mandatory requirement on them. The land registration authority only registers the deed and does not issue a certificate of rights to the landowner.

(2) **The rights registration system.** Land registration under this system is an effective, essential requirement in any change of land rights, and any acquisition or change of land rights is ineffective without registration. The land registration authority not only examines whether the documentary materials submitted by the applicants meet the formal requirements, but also conducts a substantive examination on the reasons and facts of the change in land rights and whether it complies with relevant laws and regulations. The rights to land become effective as soon as they are registered. All occurrences and changes in land rights must be registered. The registration authority records these things

only in the register and does not issue land rights certificates to those having rights to the land.

(3) **The Torrens registration system.** Under this system, it is not mandatory to have all land registered, but once a parcel of land is registered for the first time, then each subsequent transfer or change of land rights to that parcel of land must be registered and will not be effective unless it is registered. Once the land rights are registered, they cannot be overturned—an effect that is guaranteed by the state. The state specifically draws a certain percentage of the collected land registration fees as a security deposit, which is used to compensate landowners for losses caused by faulty registration. Two copies of the land register are filled, with the original retained by the registration authority and a copy obtained by the rights holder. The contents of the original and the copy are identical, and the copy serves as the land rights certificate held by the landowner.

Third, building a land market with equal access and fair trade.

(a) The first thing to do would be to build a system in which the rights appertaining to landownership of both kinds are equal. It is necessary to change the current system in which the rights appertaining to ownership of the same parcel of land differ depending on which ownership system is being used, and to extend the same rights of occupation, use, income and disposal to land whether it is under state or collective ownership. By giving equal rights protections for both kinds of landownership, provisions of the *Constitution* and relevant laws guaranteeing the same rights for the same land would be realized. The use regulation system should be the only control on access. Under the use regulation system, farmers' collective land and land under other entities would enjoy an equal right to access nonagricultural use and an equal right to the value-added income of the land. The main function of a clear and definite plan is to bring into reality certain spatial and functional layouts. We need to change the pattern in which local governments use planning revisions to convert farmers' collectively owned land into state-owned land and government-managed land. We also need to do away with the current policy in which a line is drawn to divide the city from the village, as a result of which there are different kinds of access for land inside and outside the city circle, depending on their ownership. Excluding land outside the circle that may be used for non-profit, nonagricultural construction, land within the city circle that is owned by farmers' collectives should also be used for nonagricultural construction without having to change the nature of its ownership, provided that the construction is in compliance with use controls.

(b) It will be necessary to clearly limit state-ownership of urban land to land held in store in built-up areas. It should be possible for collective ownership to be retained where newly-added construction land is being used for nonagricultural economic construction; land expropriation is to be the exception, where its purpose is for the public interest. As for collectively-owned land in built-up areas, the methods of "protecting rights and distributing the interests" or "transferring rights and protecting the interests" may be adopted to protect farmer's land property rights.

(c) The land market should be rebuilt to handle equal transactions. This would involve establishing an open, fair, equitable, and unified trading platform and trading rules that would bring such a market into reality—a market where the participating entities are treated equally and where their transactions determine supply, demand, and price. We must break the current pattern in which local governments have an exclusive monopoly of the land supply. As long as the participating entities—be they the government, farmers' collectives, or units using state-owned land—comply with the relevant laws and trading rules, they should be able to engage in land transactions in a unified market. The market allocation of state-owned land should be increased further through reform—gradually reducing the planned allocation portion, constantly expanding the paid use portion, and ultimately eliminating the "dual track system" of land supply. Paid use should be implemented first for land under profit-oriented management—where the infrastructure and municipal facilities exist, and there are various types of social enterprises that are competitive, income-producing, and have conditions ripe for reform. Under the unified land system, a land price system should be established, with improved secondary markets for land leasing, transfer, and mortgaging. Under the framework of collective construction land

> entering the market, "houses with limited property rights" (小产权房) that have already been formed should also be paid a certain amount of subsidy, according to their different situations, to properly resolve this problem that history has left us.

Fourth, accelerating the reform of the land expropriation system. The current system whereby compensation for land expropriation is calculated by multiples of income from original use should be corrected in a way that the compensation is fair. The resettlement system should be improved, so that land-expropriated farmers' houses within urban areas should be compensated for at market price, land-expropriated farmers' houses outside of urban areas should be compensated for at the local market price, and the farmers should be included in the urban social security systems. Various modes of resettlement with land retainment and land shareholding should be explored in order to ensure farmers' long-term livelihood. Finally, the dispute resolution system with regard to land expropriations should be improved, with relief channels smoothed out and the legitimate rights and interests of farmers with respect to their land should be safeguarded.

Box 4: Zone Expropriation in the Taiwan Region

When the Taiwan government in 1953 implemented the *Land Reform Act* (《实施耕者有其田条例》, or more literally, "Regulations Implementing the 'Land-to-the-Tillers' [Policy]"), it completely abolished the tenancy system through levies on landlords and distribution of land to tenant farmers. By 1956, the proportion of land area under cultivation of landholding farmers exceeded 85%, whereas in 1948, it had been less than 50%. After land reform, the government has provided protections for the private ownership of land by various ways and means: providing legal protections; raising legal awareness and understanding of government policies through extensive publicity campaigns; and conducting annual government-directed surveys to closely monitor the implementation of land reform. During the subsequent urbanization process, the government adopted a variety of methods to control the conversion of agricultural land to urban land, and at the same time, it helped farmers to independently develop nonagricultural uses of land.

General expropriation and zone expropriation.

Taiwan uses the enumeration method to stipulate under what circumstances private land can be expropriated. General expropriation refers to expropriation for public interest purposes, but zone expropriation refers to the collection of private cultivated land and converting it to the following uses: (a) urban development and construction; (b) rebuilding of old cities; (c) conversion of cultivated land within urban planning areas and converting it into construction land, or conversion of industrial areas into residential and commercial areas; (d) development of non-urban land; (e) rural community upgrading to improve public service facilities and public health; (f) other development purposes in accordance with relevant laws. The procedural laws and compensation laws that had to be followed in general versus zone expropriation were not the same.

In general expropriation, the compensation standard had been the land value determined by tax assessment; Taiwan's Bureau of Land administration publishes the value of cultivated land across Taiwan as determined by tax assessments every year. However, this tax assessed value standard has been replaced recently with the market value standard based on recent comparable farmland transaction prices. The procedure for determining the market value standard is relatively simple. It includes: a public hearing, a 30-day public notice for the decision on land expropriation, a written notification to the landowners affected by the land acquisition, and the submission to the local land administration of the content of any dispute over the compensation within 30 days, whereupon the land price and the compensation is reviewed by the land price evaluation committee or submitted for administrative civil suit.

Zone expropriation, on the other hand, indicates that the expropriation of the private property and its conversion to commercial use has been in fact approved, so the landowner receives higher compensation and even better procedural protections. The landowner may request monetary compensation, in which case the compensation standard is the market price of the land during its agricultural use. Alternatively, landowners may get back 40%-50% of the land as compensation. This compensatory discounted land is part of the originally expropriated farmland—but before the farmer gets it back, it has already been converted to more valuable urban construction land. As a third option, the landowner can choose part of the compensation in cash, and part in the form of discounted land.

Procedurally, the government is required to attempt negotiating the means of buying land from landowners before resorting to the land expropriation method of zone expropriation. If the negotiation fails, then the government can choose to adopt the zone expropriation method. Before expropriation, the government must conduct a public hearing in order to explain the compensation methods to the owners of the land to be expropriated. If a landowner chooses to the discounted land option for compensation, he or she has first choice on a parcel within the designated construction area. Landowners also have the opportunity to negotiate the specific area ratio of discounted land (40% to 50% of the total expropriated land area).

Financing for infrastructure in zone expropriation. The government is required to use a certain percentage of the expropriated land for public facilities, such as schools, roads, and so on. Although the law does not stipulate how large a proportion of the land must be used for public facilities, in practice usually 40% to 50% is used for public purposes. After removing the 40–50% of public facilities land and the 40–50% of land used as discounted land in compensation, the government is left with about 10–20% of the land, which it can sell to developers. To prevent irrational urbanization, the law requires the government to use the sale of the land it acquires to balance the budget of its zone expropriation project, and it prohibits the use of other local income to support the development of the zone expropriation project. Therefore, the government needs to use the net proceeds from the sale of 10–20% of the land to build all of the public facilities in the expropriated zone. After all of the land in the zone has been expropriated, it is first necessary to change the land use and register it as government-owned. After the selection of the discounted land parcels has been completed, these parcels are re-registered as privately owned land. The remaining construction land is sold to developers and is re-registered as commercial developers' land. Finally, only the land used for the construction of public facilities remains government-owned.

Fifth, establishing a system to fairly distribute land value-added income. The issue of distributing land value-added income relates to social equity, urban sustainable development, and the transformation of the usual mode of development, and it must be given top priority in land system reform. Based on the principles of generating land value-added income, we should learn from international experience and formulate principles on the rational distribution of that income. We should fundamentally reform the current situation, in which that income all goes to the government, by establishing a mechanism by which that income goes to society, that is, the original landowners are guaranteed fair compensation and differential rent on the land, cities are guaranteed the land and funds they need for urban infrastructure, and the public at large benefits. Secondly, we should make a distinction between "returning price increases to the public" and "returning price increases to the government," and prevent "the public" from becoming "the government" whenever the benefits of land added-value income appear, and even to prevent some local government officials from using it as "extra income." "Returning price increases to the public" must be guaranteed, so that the benefits of that income go back to society. Thirdly, we should learn from the experiences of Taiwan and other places in China and implement "zone expropriation," in order to promote the rational distribution of land appreciation income through land use reform. Fourthly, a land fund system should be established, wherein a certain proportion of land value-added income is collected in order to make adjustments for shortfalls and surplus, smooth out the effects of market volatility on local finances, and realize an "interannual" rational distribution of land income.

Box 5: Western Theory of Land Value Restoration

In modern times, Western scholars have done much research on the appreciation of land values. A basic consensus is that land value appreciation is highly correlated with location. In other words, the value of a parcel of land in an urbanized area is not comparable to the value of a parcel of land in a traditional agricultural area. Therefore, when a parcel of land in a traditional agricultural area is expropriated, compensation can only be made at the market price when the piece of land is being used for agriculture. Likewise, when a parcel of land in an urbanized area is expropriated, compensation must be made according to the market price of urban land. The sources of land appreciation and the distribution of land appreciation income are focuses of Western scholars' research. The sources of land appreciation can be attributed to the following factors: (1) landownership gains; (2) changes in the position of land relative to expanding towns; (3) increases in land value caused by population increase; (4) changes in land use conditions, such as volume rate increases, etc.; (5) public infrastructure building. Theoretically speaking, land appreciation income should be distributed according to its source, that is, the landowner reaps the income derived from landownership through land appreciation, and the government and society reap the income from land appreciation caused by the expansion of towns and cities and by public facilities investment. But in fact, determining how to separate land value-added income based on its causes and how to distribute that income fairly among relevant stakeholders has become a difficult problem in Western theories on land value-added income distribution.

In Western society, because of its system of private landownership, part of the land value-added income obtained by landowners is guaranteed by law. The difficulty lies in determining how to realize that part of land value-added income that belongs to society. To prevent that portion of land appreciation that was contributed by society from being enjoyed at no cost by landowners, and to return society's contribution back to society, Western scholars put forward the theory of land value capture, based on the ideas of Henry George (1839–1897). This theory holds that, although some private landowners' direct actions can also increase land value, in general, most of the value-added of land comes from the behavior of the public sector (as opposed to the private sector), for instance, through the granting of land use and plot ratio development permits, through investment in infrastructure, or through increases in urban population. In each of the above situations, the real estate owner clearly did not contribute to the appreciation of the value of the land. Therefore, the recovery by the public sector of all or part of the appreciated value of the land is a social necessity, and such recovery can also provide a source of funding for urban renewal, infrastructure building, and environmental improvements.

These scholars have also developed a set of policy tools that use land value-added income capture for urban construction and the provision of public goods. That is, under statutory planning, a part of the land under private ownership is used for the provision of related public goods, and another part of it, at a premium, as used for infrastructure building, and the land remaining with the landowner benefits from urban renewal and land value enhancement. Joseph Stiglitz, the Nobel Laureate in Economics, has summarized this as "the Henry George Theorem."

As urban planning took hold, its impact on land values and land value-added benefits increased. Urban planning means not only establishing a public order for land development by restricting the free exercise of individual property rights, but also means realizing the basic pursuit of the public interest by adjusting and redistributing the benefits of land development. To realize his "Garden City" concept, Ebenezer Howard (1850–1928) proposed that the government restrict the transfer of land use rights and retain part of the income from development as a social development fund. The well-known (in Britain) "Uthwatt Report" (1942) proposed that while urban planning will add value to some land, the interests of landowners of other land might also suffer a loss as a result, due to a loss of planning control, which would necessitate the establishment of a compensation mechanism and an interest-balancing mechanism between the two. The solution it proposed was that the government would pay the compensation amount to existing landowners, and after that the land development rights would be requisitioned uniformly. The compensation amount would be formulated by the government in order to control land speculation and reduce conflicts of interest between the government and the public.

Since the beginning of the twentieth century, due to rapid urbanization and increasing demand for public services, urban planning has had a more significant impact on the formation and distribution of development benefits, so the issue of social equity in the distribution of those development benefits has become more prominent. Western scholars have proposed a theory of Public Redistribution of Development Benefits. The term "development benefits" refers to the land value-added income caused by public investment and construction. Determining how to reasonably distribute this part of value-added income and reasonably compensate those whose interests have been harmed is at the core of the development of this theory. Its policy tool is the "beneficiary burden" (i.e., the beneficiary pays) of public projects. That is, the reasonable sharing of the cost of public investment projects and the fair distribution of benefits are linked, and the costs to be borne by the property owner are based on the extent to which he or she benefits, to realize the social restoration of development benefits.

Sixth, establishing a land property tax system befitting the development of a modern society. Land-based property tax is a kind of tax commonly levied in developed market economies that provides an important and stable source of fiscal revenue for local governments. To levy property tax on land in China would be in line with international experience; the conditions are right for it, and at the same time, it would provide a realistic basis for the reform of land finance.

There are obvious shortcomings in the current land tax system. First, with regard to acquisition and tenure, because the Urban Land Use Tax (城镇土地使用税) and Farmland Occupation Tax (耕地占用税) are levied on a quantitative basis, and the ad valorem portion of the House Property Tax (房产税) is based on a property's historical cost, the flexibility of tax collection with regard to tenure is inadequate. Furthermore, the scope of the taxation is too narrow, while the scope of exemptions and preferential treatment is too broad, so their subjective position among real estate taxes is declining. Second, with regard to land transfer, the overall tax rate of the Deed Tax (契税), Business Tax (营业税), Stamp Tax (印花税), plus taxes such as the Urban Maintenance and Construction Tax (城市维护

建设税) and Educational Surcharge (教育费附加), taken together, are too high, and they inhibit regular real estate transactions and transfers. Third, even though the Land Appreciation Tax (土地增值税) and Income Tax levied on land transfer income have shown an upward trend in recent years, they have always accounted for a very low proportion of real estate taxes, so they have failed to effectively play a role in cracking down on land speculation and improving the efficiency of the real estate market.

We propose to draw from the experiences of market economies and establish a new form of land real estate tax system that is suited to China's national resources condition. The tax system for land tenure, transfer, and income should be restructured and reformed systematically. First, with regard to land tenure, the main tasks would be to change from a unit tax to full implementation of ad valorem collection, increase the flexibility of taxation, and let real estate taxes on land fulfill their role in raising local government fiscal revenue. Second, with regard to land transfer, the overall tax rates of the various applicable taxes should be reduced, since this would help allocate resources and improve efficiency. Third, tax revenue collection and management should be strengthened, so they can fulfill their role in promoting the fair distribution of social wealth.

Seventh, establishing a state-owned land asset management system and land financing system. Government land sales should be changed to state-owned land assets management; as the owner of state-owned land, the government acquires the rights and interests to land, so it should set up a state-owned land assets company to engage in state-owned land management. Government land reserve institutions should be restructured, and a market for trading state-owned land assets should be established. The system for managing use of state-owned land assets' operating income should be improved, with clarification that such operating income cannot be used in the current period, and that its use and performance are to be monitored and reviewed by the National People's Congress. The financing system for state-owned land should also be improved, with clear warrants and terms regarding the use of land as collateral, with third-party asset assessment, and strict investigations of violations, in accordance with the law.

Eighth, establishing a modern land administration system, centered on tenure management and use regulation. A system for planning state-owned land spaces should be developed. The rigor of overall land use planning should be enhanced, and use controls should be implemented in accordance with the law. Land tenure management should be strengthened, and a unified cadastral management system established. The land index approval and annual plan management arrangement should be gradually eliminated, and in its place, central and local land management responsibility systems with equal rights and responsibilities should be established.

4. Breakthroughs and Priorities in Land System Reform

4.1 Reform Breakthroughs

In the next round of development, we cannot continue to follow the old path of outward expansion and wasteful use of land. We must instead achieve a transformation away from the total-land-supply-driven growth of the past to development through structural adjustment. Through innovations to the land system, the land use structure may be optimized, the stock of land revitalized, and the reasonable, efficient use of land ensured for the next strategic opportunity period. Through expanded pilot programs, experience for the comprehensive reform of the land system can be provided.

Specific measures would include the following:

(1) *Formulating measures on converting industrial land to commercial land, dealing with ownership management, and distributing value-added income.* Industrial land should be proportionally reduced and urban residential land increased, the urban land structure should be optimized, and industrial transformation and sustainable urban development should be promoted.

(2) *Formulating measures on land income distribution and ownership management in the conversion of public-use land to industrial land.* The proportion of public-use and especially government-use land should be reduced, and land for industrial development and livable space increased.

(3) *Formulating rules and measures on the inclusion of collective-used land within rural-urban fringe into the urban construction system.* Incorporating such collective land within these areas into urban planning will free up urban development space.

(4) *Formulating measures on making rural migrants into full urban residents and on the disposal of rural homesteads.* The structure of urban and rural construction land should be optimized. The collective construction land belonging to the rural migrant population landed in the city should be converted, gradually changing the situation of land tenure on both of the urban and the rural.

Box 6: "Revolution of the Three Olds" and Experimenting with the Land System in Guangdong Province

After rapid industrialization and urbanization, Guangdong Province has also accumulated a large amount of large amount of existing or inefficiently used for construction in store, Preliminary investigation has shown that the province has a total of 4.1 million *mu* of land for "Revolution of the Three Olds." In recent years, as Guangdong has been carrying out "Revolution of the Three Olds," it has been conducting a series of experiments to test innovations to the land system, all centering on the differences between existing and newly-added construction land in terms of occupation, development, disposal, income, and so on.

To promote the transformation of old cities, old villages, and old factories, the Guangdong Provincial Government formulated and issued "Several Opinions on Implementing "Revolution of the Three Olds" to Promoting Economical and Intensive Utilization of Land" (Yuefu 2000, No. 78; hereinafter referred to as "Provincial Government No. 78"). This embodied policy innovations and breakthroughs in five main areas:

(a) Since the cost of completing procedures for existing construction land, for which there is no legal procedure, is rather high, and the current users of that land generally lack enthusiasm for any change, we should permit the expropriation procedures of such land to be handled based on its current condition, after it has been handled (with punishments for violations) in accordance with the land management laws and policies in effect when use was granted.

(b) Because of problems such as the high cost of "Revolution of the Three Olds," the comparatively prolonged development cycle, the current single method toward rebuilding, and the difficulty of mobilizing the enthusiasm of the original land-use rights holders, the government permits negotiated sale of the land of the "three olds" as a supplementary procedure after the completion of formalities (with the exception of land being resupplied by the government after it has purchased it and held it in reserve).

(c) To protect the development income of the original land users and to mobilize their enthusiasm for supporting and participating in the rebuilding, when the government, in accordance with law, assigns construction land that it has reclaimed from business enterprises or has expropriated from collectives, in its reconstruction of old factories and villages, it gives a subsidy of no more than 60% of the net revenue of land sales to the original business enterprises or collectives that had used the land; these business enterprises or collectives then use the subsidy specifically, in accordance with relevant regulations, to support their development.

(d) To address the problems presented by urban corners (城市边角地), urban enclaves (插花地), and sandwiched parcels (夹心地), which affect the overall rebuilding and development of a land parcel, the government allows the inclusion of such corners, enclaves, and sandwiched parcels within the scope of rebuilding the "three olds," and there is no longer any need to go through the relevant approval procedures in accordance with existing laws and policies.

(e) In the process of rebuilding urban villages, rural collectives are willing to have their collective land converted into state-owned land, but the current expropriation process is complicated, and the approval process is drawn-out. Thus the government allows village collective economic organizations to voluntarily apply to the Provincial People's Government have their construction land converted to state-owned construction land.

The above policies have effectively protected the rights and interests of the original land users, fully mobilized their enthusiasm for rebuilding, and promoted smooth progress of "Revolution of the Three Olds." On this basis, a series of policy documents have been issued and a set of supporting local policies have been formulated that center on specific procedures and prices for purchasing land use rights, standards for reimbursed land prices, controls on the allocation and use of land sales income, and review and approval procedures for rebuilding projects. Together these have formed a policy system for "Revolution of the Three Olds."

The "Revolution of the Three Olds" achieved remarkable results. From 2008 to the end of 2012, 2,851 reconstruction projects over a total area of 150,000 *mu* were completed province-wide, and currently 2,920 projects are under way over an area of 200,000 *mu*. "Revolution of the Three Olds" has effectively developed the existing store of construction land; the recycling of such land was promoted, land use layout was optimized, and land use efficiency was improved. In the 150,000 *mu* of land where reconstruction has been completed, the average plot ratio has been increased from 0.66 to 1.5, and building area has increased from 66.012 million square meters before rebuilding to 153.6527 million square meters—an increase by a factor of approximately 2.3. The area of usable land increased by making room accounts for 44.65% of the area where rebuilding has been completed, saving approximately 67,300 *mu* of usable land. Based on this ratio, we can estimate that after the 4.14 million *mu* of land used by the "three olds" throughout the province has been rebuilt, more than 1 million *mu* of land can be saved, which effectively guarantees enough land to meet the needs of economic and social development.

The "Revolution of the Three Olds" has also driven social capital investment, expanded urban and rural employment, and stimulated consumer demand. Since 2008, the province has invested a total of 424.29 billion yuan in reconstruction funds, accounting for approximately 5.39% of the provinces total fixed asset investment over the same period. Of this amount, social investment accounted for 74%. As more and more social and private capital has been invested with enthusiasm into the work of rebuilding, it gradually became the predominant force for the "Revolution of the Three Olds." The "Revolution of the Three Olds" has also led to the cumulative increase of 1.99 million jobs over the past three years. The process of rebuilding and optimizing the layout of land use also promoted the adjustment of the industrial structure—it eliminated obsolete production capacity, "keeping the cage while changing the bird" (腾笼换鸟), and it promoted the development of strategic emerging industries and the expansion and strengthening of existing pillar industries. Statistics show that among the reconstruction projects, 1,898 projects involve industrial structural adjustments, accounting for 66.6% of the total. Of these, 416 belonged to the phase-out or the "two highs and one resources" (两高一资; that is, high energy-consumption, high polluting, and resource-intensive) industrial categories; at the same time, 386 projects introducing modern service industries and high-tech industries were carried out, and 298 projects involved investment of more than 100 million yuan. Realized output value (operating income) after these projects were carried out increased by a factor of 2.22 over what it had been before.

The transformation of old villages also promoted rural infrastructure building, thus changing the villages' dirty, shabby condition, backward infrastructure, and building safety issues. Statistics reveal that 866 of the completed rebuilding projects were to build urban infrastructure and public facilities, involving 42,000 *mu* of land. More than 6,600 *mu* of public green space was also added. During the rebuilding process, the land that was freed up was used to develop the collective economy, which increased rural collectives' income. According to statistics, the old villages' collective income has grown by a factor of 2.5 from the time before reconstruction.

4.2 Reform Priorities

Considering how even a single reform in the land system eventually affects the whole, we recommend that a few typical regions be selected first for experimentation, and upon summarizing the results of these experiments, the reforms may be elevated to national policy. At the same time, it will be necessary to develop a reform agenda and sequence of priorities.

In the first stage (2013–2014), on the basis of a summary of existing pilot projects in land reform, the scope of the pilots should be expanded. Registration and the issuance of certificates confirming rights to contracted land, homesteads, forest land, and so on should be completed. Reforms to optimize land use structure and promote market entry of existing construction land should be advanced, as well as a reform of the land expropriation system that has fair compensation at its core. Local financing platforms should be regulated and the scope of local bond issuance expanded.

In the second stage (2015–2017), the focus should be on forming a unified urban and rural land market, exploring the state-owned land capitalized management model, establishing a land property tax system, forming a new land value-added income distribution mechanism, and forming local government financing system with diverse revenue streams.

In the third stage (2018–2020), a modern land system, wherein ownership of the two kinds of land entails the same rights, and which is managed through planning and use regulations, should be basically established.

8

Land System Reform and Its Implementation as Decided in the Third Plenary of the 18th CPC Central Committee

Judging from the "Decision of the Central Committee of the CPC on Some Major Issues Concerning Comprehensively Deepening the Reform" (《中共中央关于全面深化改革若干重大问题的决定》), approved in the Third Plenary Session of the 18th Central Committee (November 2013), the land system reform this time will be a comprehensive, integrated, and systematic one. Almost all of the aspects of land-related issues are involved in the reform program. That is, it involves the future institutional arrangements for land rights and for agricultural management in rural areas, during the transition from a rural society to an urban one. It involves the formation and development of the land market during the conversion of agricultural land to urban and industrial land, as well as the reform of the land expropriation system. It involves the most critical issue of interest in the urbanization process, i.e., the fair distribution of land value-added benefits among stakeholders. It involves the structural transformation of land in the non-agricultural land market. Finally, it involves the building of the territory space system and the future design of the framework of the land management system.

1. The Relationship Between Land System Reform and Overall Reform

First, how well land issues are resolved will be a litmus test for the modernization of national governance. The "Decision" proposes advancing the modernization of state governance in terms of its systems and capabilities. The implications of this important expression still need further research and exploration. We believe that the reform should include a set of systems and organizational structures that will modernize the country that form a fair and just institutional environment, one ruled by law, and ensure that the vast majority of ordinary people will be able to share the fruits of modernization. Land is a vehicle that affects the interest relations of the greatest number of stakeholders, and everyone living in our society is directly or indirectly connected with it. We are currently in a special period, when land issues are most complicated and difficult. It involves the land issues of our traditional agricultural society where land allocation and efficiency, political stability, and agricultural product supply should be taken into account. It also involves the land issues of society in transition, so it must directly face the issue of the source of that huge value-added revenue generated when agricultural use is converted to urban use, and the issue of how to distribute it fairly.

If the land issues are not resolved well, other problems in governance will emerge, and then it will become impossible to speak of "modernized governance." But if the land issues are resolved well, then most issues relating to the unfair distribution of interests and revenues can be solved to a certain extent.

Second, the method of allocating land resources will be changed, so that the market plays an important, decisive part in it. The market's decision role here was the main line of the Third Plenary Session of the 18th Central Committee. The "Decision" regards "establishing a unified market for urban and rural construction land" as an important part of "building a unified, open, competitive, and orderly market system," thus highlighting the importance of deepening land marketization reform. The biggest land problem we are facing lies in the government-led allocation of land resources. For example, when agricultural land is converted, the government always does it by land expropriation, after which the government becomes the sole allocator of the land, supplying it for use by industries and cities. Much of this land is still supplied through planning (not in exchange for payment). In other words, China land market is dominated by local governments, so that supply and demand are severely distorted.

If the government continues to dominate the allocation of land resources and does not let the market play its role, a series of problems will arise. For example, land resources are being used inefficiently, despite China's scarcity of land, but everywhere one looks there are garden factories (花园工厂), and there are a lot of large squares, broad highways, and large office buildings in the city. This is a sign that the land use structure is distorted and the proportion of industrial land is too high, so that land for residential use and service industries becomes crowded. The distorted land market has led to distorted real estate markets and distorted economic structures. Shifting leadership of land resource allocation from the government to the market will be conducive for improving land use efficiency, adjusting the economic structure, and transforming development methods.

After the market's decisive role in land resource allocation is realized, the government can also play its correct role. Planning of territory space, managing land property rights and cadaster, ensuring fair distribution of land interests, and controlling land use are all things that the government should do. With regard to land, clarification of the relationship between the market and the government can make the land market's development healthier and more complete, and economic development will become more sustainable. The construction of the ecological civilization, the protection of cultivated land, food security—all these and more depend on the government's correct administration of land. Therefore, the proper handling of the relationship between the government and the market with respect to land will decide whether the main line of the "Decision"—that the market plays a decisive role—can be realized.

Third, reforming the land system is the key to integrating urban and rural development. In the "Decision," improving the institutional mechanisms of the integration of urban and rural development is the key to integrating that development. Since the 16th Party Congress (2002), the coordination of urban and rural policy has played a significant role in promoting the balanced allocation of public resources and public services between urban and rural areas. The "integration of urban and rural development" proposed by the "Decision" has at its core the achievement of shared development and prosperity between cities and rural areas. The key will be to let production factors flow reasonably between urban and rural areas. That is, we cannot allow all resources to flow to the cities and none to the rural areas—the villages cannot live on blood transfusions alone.

The land system holds the key to realizing a reasonable flow and equal exchange of production factors between urban and rural areas. In our current land system, the process of urbanization consists of nationalizing land, and farmers losing their land. If this sort of land system is not reformed, the ultimate result of urban-rural integration can only be that the cities swallow up all of the rural areas and cities develop while the rural areas wither.

2. Establishing a Unified Market for Urban and Rural Construction Land

The Third Plenary Session of the 17th CPC Central Committee (2008) once proposed "progressively establishing a unified market for urban and rural construction land," and the pathway for achieving that would be to permit rural collective construction land to be used for purposes other than agriculture and to be spared from expropriation, provided that the land lay outside of the urban planning area (outside the city circle), was not used for public facilities. However, this policy could not be implemented due to the small scale of construction land outside of the city circles, plus the fact that the expropriation of land beyond the city circles continued unabated. Meanwhile, since land in the urban area (within the city circle) was state-owned according to the law, the only way to obtain more land in the urban area was to continue expropriating it. Therefore, even though the Third Plenary Session of the 17th CPC Central Committee proposed "progressively establishing a unified market for urban and rural construction land," it was in fact very hard to advance.

The Third Plenary Session of the 18th CPC Central Committee not only rescinded the word "progressively" from the 17th CPC Central Committee's statement, but also formulated an institutional arrangement wherein collective construction land within a certain scope entering the

First, the "Decision" proposes to permit rural collective commercial construction land to have the same market entry, pricing, and rights of sales, leasing, shareholding, and so on as state-owned land, provided that the land is being used in compliance with planning and use regulation.

First, it defines the scope of the marketization of rural collective land, to wit, collective commercial construction land. China's current laws stipulate three kinds of collective construction land: (1) land used by business enterprises set up and operated by farmers' collectives or by business enterprises that they set up and operate jointly with other units or individuals through shareholding of land use rights, joint venture, and so on; (2) land used for collective public facilities and public benefit facilities; and (3) rural households' homestead. According to existing laws and regulations, the first category obviously belongs to "collective commercial construction land" referred to in the "Decision," but the second category obviously does not. The third category, however, requires a little discussion.

Based on the *Law on Land Administration*, rural households' homesteads are obtained by farmers as members of their collectives' economic organizations; each farmer's household receives one homestead, and the area standards are determined by each province, autonomous region, and municipality. If rural residents sell the houses they live in and then apply for another homestead, the application will not be approved. The *Property Law* stipulates that a homestead is a usufruct, which generally means that "the owner of the usufruct shall, within the extent permitted by law, enjoy the rights to possess, utilize, and obtain profits from the immovable or movable properties owned by others." The provision regarding the right to the homestead, meanwhile, states, "The owner of the right to the use of a homestead shall enjoy the right to possess and utilize land owned by a collective, and the right to construct residential houses and their auxiliary facilities on such land, according to law." Thus, in comparison with the usufruct in its general sense, the usufruct to the homestead lacks the "right to obtain profits from" it, and the law also specifies clearly that "such laws as the *Law on Land Administration* and relevant state regulations shall be applicable to the acquirement, exercise, and transfer of the right to use a homestead." In other words, based on current laws, the right to a homestead is mainly the right of residence, so a homestead does not belong to the "collective commercial construction land" referred to in the "Decision."

Second, this is the first time that the concept that the "same rights appertain to lands for the same use" (同地同权) has been written into one of the Party's highest documents. The biggest drawback of our country's land system has been that different rights appertain to the same land, depending on whether it is collective or state-owned, but the "Decision" clearly allows rural collective commercial construction land to enter the market on an equal basis with that of state-owned land. We believe this should include the following levels of meaning: (1) That the "same rights appertain to the same land." That is, collective commercial construction land of the collectives may, just like state-owned land, be assigned, leased out, and held as shares; this would break the barriers caused by forbidding collective construction land to be leased out. (2) That land can enter the market on an equal basis. That is, collective commercial construction land and state-owned construction land can legally enter into the market on the same platform. This would change the current situation in which state-owned construction land is exclusively traded on the platform, and collective construction land is traded in unregulated fashion on the grey market. (3) That prices are determined by supply and demand. Bringing collective commercial construction land into the market will alter the current situation, in which the government has an exclusive power to assign use rights and thereby control supply, which results in price distortions. When a number of collective economic organizations enter the market with their collective commercial construction land, it will really form a mechanism that uses supply and demand to determine price, and it will promote the healthy development of the land market.

Third, planning and use regulation must be complied with. Farmers' collectives can use their collective commercial construction land for nonagricultural economic activities only if they comply with planning and use controls. Therefore, the "Decision" places strict restrictions on the entry of collective commercial construction land into the market. The first restriction is that it makes only a small opening for collective commercial construction land—it is not the case that all collective land enters the market at the same prices and with the same appertaining rights as state-owned land. The second restriction is that even this collective commercial land must be subject to planning and use regulation.

We must squarely face the fact that collective commercial construction land entering the market will encounter huge challenges. This is due mainly to the fact that a large quantity of collective construction land has already entered the market, and most of this by far goes beyond what current laws define as collective commercial land. First, at present, a large percentage of the existing collective construction land in coastal regions such as the Pearl River Delta (珠三角), the Yangtze River Delta（长三角）, and the Bohai Rim(环渤海地区) has already been used to develop industry. Here, it includes not only what the current law defines as collective commercial land, but also land used for collective public facilities and public welfare, and a large quantity of farmers' homesteads. Second, this collective construction land that has already entered the market exists in a gray state. Most of the collective construction land leased to business enterprises by village economic organizations use land lease contracts that have been signed by the two parties in private, and in the event of a dispute, the court treats the contract as invalid, resulting in losses to both parties. Third, in the urban-rural fringes within urban planning areas, there exists a large quantity of collective construction land that is being used to operate businesses—which is at odds with the law, which states that urban land is state-owned. Thus, determining how to implement the spirit of the "Decision" and how to solve the problems of existing collective construction land that is already being used by businesses with the courage to reform according to the problem-oriented approach are urgent issues that need to be resolved in the next step of implementation.

Second, the "Decision" proposes to reduce the scope of land expropriation, standardize the land expropriation procedures, and improve the protective mechanisms for land-expropriated farmers by making those mechanisms reasonable, standardized, and diverse. If collective construction land is going to enter the market, we must be determined to resolve the problems of land expropriation. Since the 16th CPC National Congress (2002), the party and government have worked to promote the reform of the land expropriation system, and they have encountered may obstacles, the most important of which is that the traditional development methods supported by this system are difficult to change and overly dependent on it. However, due to the growing social conflicts in the operation of this system and the ever-increasing costs of land expropriation and demolition, it is now time to change.

The first task is to reduce the scope of land expropriation. According to the principles of the *Constitution*, land can be expropriated only to serve the public interest. However, because it is difficult to define which uses serve the public interest, it has been impossible to produce a catalog of those uses. We recommend that rather than negotiating and never resolving the issue, it would be better to formulate a negative list for land expropriation, as is done with reform in other areas. That is, if the purpose of use is to make a profit, then the land cannot be expropriated.

The second task is to standardize the land expropriation procedure. This item was proposed long ago, but it has been poorly implemented. The main reason is that under the current system, the two parties to the expropriation cannot come to any agreement about compensation prices and resettlement methods. In the end, it is often the case that informal solutions are adopted, which are neither transparent nor fair. We recommend that after the scope of land expropriation and the guarantee mechanism have been established, an advance procedure should be added to the land expropriation procedure, permitting an advance negotiation between the future land user and the farmers' collective, and then, if the negotiation reaches an impasse, the land expropriation process can begin.

The third task is to establish a set of protective mechanisms for land-expropriated farmers that are reasonable, standardized, and diverse. The establishment of these mechanisms is at the core of land expropriation reform, and its content in terms of policies and laws should be clarified as soon as possible. Our recommendations are as follows. "Reasonable" would be in reference to compensation price. That is, under the principle of use control, market-based compensation for expropriated land should be implemented in accordance with local conditions. "Standardized" would be in reference to land expropriation actions.

After the scope of land expropriation is clarified, and the compensation prices and procedures are standardized. If land expropriation must be carried out, it must be based on law—instead of the way it is done now, where in many cases "cash settlements are used to smooth over conflicts," or the compensation prices are raised to a very high level in certain locations and regions, so that honest people suffer. "Diverse" would refer to handling resettlement in a diversified manner. We cannot give land-expropriated farmers no more than a "kick" based on money settlement alone. If we really want to effectively resolve the issue of the long-term livelihood of farmers who have lost their land, a substantive institutional arrangement here would be to protect the land development rights of land-expropriated farmers. That is, when land is expropriated from farmers, a part of that land should be preserved for them and included in the urban planning. And this part of the land should be used for nonagricultural construction without changing the nature of collective ownership, it is the fundamental solution to the issue of the long-term livelihood of land-expropriated farmers.

Third, the "Decision" proposes to expand the scope of paid use of state-owned land and to reduce allocation of land not used for the public benefit. One of the shortcomings of the current system of land resources allocation in China is that after the government acquires land, a large part of it gets allocated through nonmarket means. The nonmarket allocation of land used for infrastructure, government, municipal facilities, and so on has resulted in the building of huge squares, wide roads, luxury office buildings, etc. In the rush to build infrastructure, it has occurred that a large amount of land allocated for the public interest has also been used for commercial purposes, so that the revenue from the land is unfairly dominated by the nonpublic sector. To improve the efficiency of land use and the fair distribution of land income, expanding the scope of paid use of state-owned land and reducing the allocation of nonpublic land is necessary. Paid use should be implemented first for the competitive commercial land with operating income and mature reform conditions in the infrastructure, municipal facilities and various types of social enterprise. By the time the 12th Five Year Plan ends (2015), the use of all types of construction land will be paid for, except for military, affordable housing, and other special uses involving state security and social public order.

Fourth, the "Decision" proposes to improve the secondary market for the lease, transfer, and mortgage of land. Under China's current land system, even though the government monopolizes the primary land market, after it assigns state-owned construction land to users, the secondary markets for land lease, transfer, and mortgage have not only failed to develop, but continue to weaken. The outcome is that once land use has been assigned, whoever gets the land value-added revenue is the one who benefits. Among the various kinds of existing state-owned construction land, one kind in the past was occupied by various units and enterprises by allocation—that is, without having paid for it—and another kind was acquired by enterprises through payment. In both kinds, however, all of the land value-added revenue has been obtained by the current owners, with the outcome that the value-added revenue of state-owned land has been unfairly distributed, and it leads to the loss of land value-added revenue. To improve the secondary market for land leasing, transfer, and mortgage, we suggest that relevant departments should introduce regulations to allow existing state-owned construction land to enter the open market and be traded there, and to improve transaction rules, formulate income distribution principles and related tax measures. All these would promote greater activity and development in the secondary land market.

Fifth, the "Decision" proposes the establishment of an effective, reasonable price comparison mechanism for industrial land and residential land, and to increase the price of industrial land. This reform aims to address the irrationality of the price structure of construction land in China. In the last round of development, large quantities of scarce land were supplied as industrial land at low prices in order to promote rapid industrialization, resulting in an excessively high proportion of industrial land and excessively low proportion of residential and service industry land, so that livable urban spaces got crowded, the real estate market was distorted, and the development of service industries was hindered. In the next round of development, the land-use structure must be optimized so that the relationship between industrialization and urbanization is better matched. The core of this process is to change the current method for allocating industrial land and the price formation mechanism, so that the price of industrial land is truly determined by the market and the prices of industrial land relative to residential land become more reasonable. This will be an important institutional arrangement for the transformation of the development mode and a focus when the next reform of the existing land use system is carried out.

3. Giving Farmers More Property Rights

The ultimate goal of reforming the land system is to increase farmers' land property rights by empowering them more fully with rights to their land. Therefore, a sign of whether or not the land system reform of the Third Plenary Session of the

18th Central Committee is successful will be whether its promise to "give farmers more property rights" can truly be fulfilled.

First, the "Decision" proposes to give farmers the right to tenure, use, revenue, transfer their contracted land and to mortgage and guarantee their land contractual management rights; it also allows farmers to use their land contractual management rights to invest in shares to develop agricultural industrialized operations. The land contractual operating rights system is the most basic institutional arrangement of the rural land system. It has placed an important role in increasing farmers' enthusiasm for production and maintaining rural stability. Farmers' land contractual management rights are also guaranteed by law. However, in the current system of land contractual operating rights, the protection of contracting rights and management rights is combined. When a farming household is that of an owner-farmer (自耕农), the operator is also the owner of the contracting right, and there is no problem with his rights protections. However, as more and more rural laborers leave agriculture, their contracting rights and management rights are actually separated, so that the institutional arrangement in which these two rights are combined has revealed obvious shortcomings. Whenever the grassroots governments and collective organizations strengthen transference of management rights, it often leads to the loss of farmers' contracting rights, but at the other extreme, when contracting households are worried about losing their land contracting rights, they can only transfer them within the family, to relatives, or to a close acquaintance in their home village, which affects the efficiency of land management.

For these reasons, this "Decision" makes a policy separation between contracting rights and management rights and empowers farmers these rights separately. That is, contracting households enjoy the rights to tenure, use, revenue, and transfer of their contracting rights, and managing households who have operating rights to land possess mortgage and guarantee rights. In this way, the reform is beneficial not only for the contractor, who can transfer the contracted land safely, but also for the manager, who can use the management right as collateral and guarantee, obtain financial support, and engage in modern agriculture.

As for "allow[ing] farmers to use their land contractual management rights to invest in shares to develop agricultural industrialized operations," this is also a policy permission for the construction of farmland stock cooperatives in some areas recently. The basic method is that farmers who go out to work elsewhere convert their contracting rights into shares to form and establish a stock cooperative, which then entrusts the land it has thus gathered to other entities that manage the land. Then, after the cooperative and the entrusted party divide the shares, the cooperative distributes the dividends to its members based on the shares each holds. This not only solves the problem of farming plots that are excessively scattered and too small, but also preserves farmers' land contracting rights, promotes moderate scale operations on the land, and realizes agricultural industrialization.

Second, the "Decision" proposes to protect the rights of the members of farmers' collective economic organizations, actively develop farmers' cooperation in shareholding, and grant them the rights to possess, derive income from, receive paid compensation upon withdrawal from, mortgage, guarantee, and inherit collective assets. There are three levels of meaning here. The first is to reform collective ownership as currently practiced, clarify the structure of collective ownership by placing members at the core. It is necessary to reform the current situation wherein collective ownership is reduced to ownership by a small number of cadres, and clarify the rights of members of farmers' collective organizations. This means that as long as a farmer is a legal member of a collective organization, he has equal property rights over the collective's assets. The second consists of adopting a joint-stock cooperative system for the management of collective assets. The assets of each collective member are quantified into shares, and based on the principle of the joint-stock cooperative system, the collective asset use, management, and distribution of benefits is carried out. This changes the current situation wherein a small number of cadres dominate and actually control collective assets, so that most members of the village collective organization cannot share the benefits. The third is to make collective asset shares fully empowering. In the "Decision," the rights of possession and income are attached to collective asset shares, and the members of collective organizations are given the rights to withdrawal and inheritance as well. More meaningfully, the mortgage and security rights appertaining to collective asset shares are clearly granted. By making collective asset shares fully empowering, institutional guarantees for the reform of the collective assets system are provided, and this will promote the management, development, and growth of collective assets.

Third, the "Decision" proposes to protect farmers' usufruct on their homesteads and reform and improve the rural homestead system; it also selects a number of trial programs in which the rights of farmers to mortgage, guarantee, and transfer their residential property rights are prudently and steadily advanced, and in which channels for farmers to increase their property income are explored. We may analyze these proposals, in order, as follows.

The first item is protecting rural households' homestead usufruct. The bottom line of the reform of the rural homestead system is ensuring the farmers' rights to tenure and use of those homesteads. The ownership of the homestead belongs to the collective, and the use rights belong to the household of the farmer living in the homestead. However, in reality, the collective organizations often have more actual and greater control of the possession of homesteads than the farmers do, and the emphasis in law and policy on collective ownership of homesteads is obviously stronger that the emphasis on the farmers' use rights on the homesteads. This easily leads to infringement on the farmers' use rights on the homesteads. We suggest that amendments to the relevant laws should follow the principle of farmers having the usufruct on the homesteads, which at present are not clearly specified and emphasized. On this basis, in accordance with general usufruct principles, the usufruct on homesteads would be improved, so that the current possessors of homesteads are granted the rights to derive income from, and transfer their use rights of homesteads.

The second item is reforming and improving the rural homestead system. At present, as more of the rural population leave the villages, the rural homestead system shows a lack of adaptability, which is manifested mainly by the fact that the free acquisition of homesteads is not conducive to intensive land use or the protection of cultivated land. Requiring membership status of people with homestead is incompatible with the fact that the boundaries of many villages are open, and people live in mixed situations—there is a contradiction between the nature of residence in a homestead and the property rights. We suggest that in the spirit of the Third Plenary Session, different types of regions should be selected for the pilot programs to reform the homestead system. The focus should be on exploring the paid acquisition and use of homestead, homestead exchanges and circulation, and breaking the boundaries between homestead members and village communities. It should be oriented basically on gradually realizing the exchange of the right to welfare for property rights, realizing farmers' complete usufruct on homestead, and promoting the effective use of land.

The third item is selecting a number of trial programs in which the rights of farmers to mortgage, guarantee, and transfer their residential property rights are prudently and steadily advanced. Judging from its wording and choice of objects, the "Decision" passage grants those rights to the farmer's housing, not the homestead, and it reasonable under the constraints of the current system and laws. As we have analyzed it above, the current homestead system consists of a combination of three rights: ownership right to the homestead belongs to the collective; use right belongs to the farmer; and ownership of the farmhouse belongs to the farmer. Since ownership rights and use rights of a homestead are of different strengths, if the rights to mortgage, guarantee, and transfer homesteads are granted, then in practice it is quite likely that the homesteads will be used by the collective organizations to collateralize loans, use as guarantees, and transfer: the collective organizations would instead infringe on the farmers' use rights to their homesteads. By contrast, the farmers' houses are their private property and have always been protected by law. If they are clearly granted the right to mortgage, guarantee, and transfer their houses, it would be conducive for them to realize more securely their income on property. Of course, at first the granting of such rights can only be done in a few pilot areas. After conditions are ripe, then house site mortgages, guarantees, and transfers can be carried out.

Fourth, the "Decision" proposes the establishment of a market for the exchange and transfer of rural property rights that operates in an open, fair, and regulated way. Currently, with regard to the issue of collective assets, the vast majority of farmers have never received any benefits from rising collective property values, mainly because a small number of village cadres have been privately disposing of and trading those assets in an unregulated manner. Establishing a market for exchanges and transfers of rural property can provide a legal channel for realizing property values of the rural collective assets. It will ensure that no group of people will swallow up all of the property value of rural property rights. When it becomes possible to trade collective assets in the light of day and members of collective organizations can share asset value fairly, this marketization of rural assets and reform of rural governance will bring revolutionary changes. The pilot programs to build rural property rights transfer and exchange markets are already under way in some localities, and the effects are quite obvious. When expanding implementation in the next step, several points will need to be clarified: (1) whether all rural collective assets are to be traded on a single platform, or whether they are to be traded in separate markets, depending on their different asset characteristics and management focal points; (2) which level of market is to trade and exchange rural assets—the city, county, or village?; and (3) the necessity of formulating management measures for collective asset transfers as soon as possible in order to guide local advancement and prevent it from going astray.

Fifth, the "Decision" proposes the establishment of a mechanism for the allocation of land value-added revenue that takes equally into account the interests of the state, collectives, and individuals and reasonably increases individual revenue. This is a very important part of increasing farmers' property-derived income. The reform is oriented toward changing the present situation, in which land value-added revenue is first obtained by the government when it assigns land and later obtained exclusively by whoever has received that land, while the farmers receive no share at all; the reform aims to increase the share going to farmers. At the same time, the "Decision" is very clever in its wording. Regarding the distribution of the land value-added revenue to the state, collectives, and individuals, it uses the word for "takes equally into account" (*jiangu* 兼顾), and the word "reasonable" (*heli* 合理) is applied to the increase in individual revenue. The main reason for this is that income from land appreciation is a very complex affair. It cannot be reserved exclusively for the government, nor can it be earned exclusively by any group or individual, and of course it cannot all go back to the original landowner. We believe that the reasonable and fair distribution of land benefits should follow the principles by which land value-added revenue is generated. From a practical and realistic perspective, land value-added revenue is generated through multiple factors, including planning, use control, infrastructure investment, and the agglomeration of industries and people that occurs during the urbanization process—and, of course, it also includes the rights and interests of landownership. The fair distribution of land value-added revenue should comprehensively consider these major factors, and on this basis, develop fair and reasonable distribution ratios.

4. Building a New Agricultural Management System

Within the rural land system, one of the most important institutional arrangements is the agricultural management system. In fact, the most important achievement of China's rural reforms has been the establishment of the latter system, which is family-based and "combines unified management with independent management." However, in the past thirty-plus years of reform, there have been huge controversies regarding the status of family-based management and the relationship between unified management and independent management. In the context of rapid industrialization and urbanization, as the agricultural labor force is declining and agricultural development methods are undergoing a major historical transformation, how we build an agricultural management system suited to the new situation will determine our choice of route for China's agricultural modernization. The "Decision" devotes a great deal of space on plans to accelerate the construction of a new kind of agricultural management system, and it is clear that this issue had received considerable attention.

First, the "Decision" proposes to adhere to the collective ownership of rural land, protect farmers' land contractual management rights in accordance with law, develop and strengthen the collective economy, and stabilize rural land contractual relations and retain them over a long term without change. The arrangement of rights pertaining to agricultural land is a precondition for dealing with the institutional arrangements for agricultural management. The great success of China's agricultural land system reform has been attributed to the separation of agricultural landownership from the use rights, and then the continuous improvement of agricultural land property rights and the stabilization of farmers' land contracting expectations while firmly maintaining collective ownership. The "Decision" of the Third Plenary Session of the Eighteenth Central Committee reaffirms the current structure of agricultural land rights, that is, while maintaining the collective ownership of rural land, it safeguards farmers' land contractual management rights in accordance with the law. At the same time, it reaffirms "stabilizing rural land contractual relations and retaining them over a long term without change" proposed in the Third Plenary Session of the Seventeenth CPC Central Committee (2008). Next, it calls for the realization of institutional arrangements for "long term without change," thus effectively resolving the problems with long-term stability and farmers' expectations relating to their land contractual relations.

Second, the "Decision" proposes to uphold the basic status of the family's management in agriculture and to promote innovation in the joint development of methods of agricultural management, including family management, collective management, cooperative management, and enterprise management. Innovation in agricultural management methods is at the core of the agricultural management system. On the one hand, the "Decision" states clearly that we must "uphold the basic status of the family's management in agriculture," which is a rational choice based on China's agricultural characteristics and international agricultural development experience. The positive and negative experiences from China's long history of agricultural development and from institutional changes since the establishment of the New China teach us that family management is the most appropriate organizational approach in Chinese agricultural operations. Experiences in international agricultural development also show that family management is the main form of organization also in the large-scale agriculture of Europe and the United States and small-scale agriculture in Asia, which indicates that family management is not directly related to acreage. The reason why family management is the main option in agriculture is determined mostly by the industry's characteristics. One is that it requires the operators to diligently take good care of nature, farmwork, and all of the many aspects of farm operation, and it is not easily standardized, unlike factory systems. Second, once workers are hired in agricultural production activities, operators need to face high supervision costs, and the efficiency of its operation is eroded.

Meanwhile, in light of the current changes in China's agricultural development and in people's relationship to the land, the "Decision" proposes "the joint development of methods of agricultural management, including family management, collective management, cooperative management, and enterprise management," which is a choice that keeps pace with innovative managerial approaches in agricultural operations at the present stage. More diversified management methods, including family management, collective management, cooperative management, and enterprise management, provide a basis for enhancing the selectivity of these methods, and are conducive for improving the methods applied for different crops, regions, and phases of development. They make production relationships more adaptable for the improvement of productivity.

Third, the "Decision" encourages the transfer of contractual management rights in the open market toward large professional households, family farms, farmers' cooperatives, and agricultural enterprises, thus developing diverse forms of operations of scale. With the above-mentioned innovations about agricultural management methods, it is logical that land contractual management rights should be transferable to all of these various operating entities, that is, the large professional households, family farms, farmers' cooperatives, and agricultural enterprises. The purpose is to achieve operations of appropriate scale. It should be emphasized that the transfer of land should be carried out in an open market.

Fourth, the "Decision" proposes to encourage the development of cooperative economy in rural areas, support the development of large-scale, professional, and modern operations, allow fiscal project funds to be directly invested in eligible cooperatives, allow assets in the form of fiscal subsidies to be transferred to cooperatives for their holding and management, and to all cooperatives to launch and develop credit cooperation. This article concerns the future development of cooperative economy and clarifies the direction of future support for cooperatives. The aim is to strengthen and expand cooperatives, so that in the future they will become vehicles for bringing farmers to the market and increase the degree of agricultural organization.

Fifth, the "Decision" proposes to encourage and guide industrial and commercial capital to villages for the development of modern seed and breeding industries that are suitable for business enterprise management, and to import modern production factors and business models into agriculture. This changes the original policy and regulations somewhat. In the past, industrial and commercial capital was not encouraged to go into the countryside. This time, it is encouraged and guided, but there are still several restrictions on the destination of that capital. One is that it should engage in projects suitable for management by business enterprises, and the other is a direction to engage in the modern seed and breeding industry.

However, in the future, how industrial and commercial capital will play its role in agriculture, and where it really will be used, needs to be further explored.

5. Strategies for Reforming the Land System

First, advance proactively and steadily, according to unified arrangements by the Central Committee of the Communist Party of China. The "Decision" proposes the "establishment of a unified market for urban and rural construction-use land," which clarifies the direction, focus, and requirements for deepening the reform of the land management system. The Economic Work Conference of the Central Committee of the Communist Party of China (中央经济工作会议), the Urbanization Work Conference of the Central Committee of the Communist Party of China (中央城镇化工作会议), and the Rural Work Conference of the Central Committee of the Communist Party of China (中央农村工作会议) made clearer arrangements for the reform of the land management system. Carrying out the reform of the land management system must be proactive and purposeful on the one hand and cautious and steady on the other.

On the one hand, we must proactively promote and improve the current land administration system. Due to the shortcomings of that system, the pattern of outward growth and expansion has acquired an inertia of its own, which restricts the transformation of economic development methods, and the nonmarket allocation and price distortion of some land causes a lot of that scarce land to be used too freely and uneconomically. The land use structure is unreasonable, and in some places "using land to plot development," to a certain extent, has created local government debt and latent financial risk. The unfair distribution of land interests and land-related disputes trigger social problems. Thus in promoting and implementing reform, we must be proactive and problem-oriented—solving the deep-rooted conflicts and problems in the land sector and reducing the unhealthy, uncoordinated, and unsustainable development of the economy and society.

On the other hand, we must proceed cautiously and steadily with the reform of the land management system. The land management system is the basic system of China's most basic political and economic system—the socialist system with Chinese characteristics, which is the prerequisite for any reform. Since the reform concerns the "big picture" of China's economic and social development, we must evaluate the possible impacts of the changes in the land system arrangements on its overall operation and its risks. The land issues that our country is facing are very complicated and affect the agricultural, transitional, and urban sectors of our society—there are too many stakeholders. The theoretical preparation that China has made in approaching the land administration system reform are quite insufficient, so in proceeding with it we must give balanced consideration to reality, policy, and law. Therefore, to ensure the reform's success proceed cautiously and steadily, with a focus on strategy.

When General Secretary Xi Jinping was doing an inspection in Shandong, he emphasized the necessity of correctly promoting reform and insisted that reform is what improves and develops the socialist system. We must accurately advance reform and conscientiously carry out the demands of the central Committee, and we must not blindly advance things without understanding them first. To advance the reform in an orderly fashion, we should not rush ahead of what the central Committee' unified arrangements, nor delay progress as soon as possible, nor rush ahead of what the pilot programs are doing, nor push ahead without prior research out of the anxiety to succeed, nor leaping ahead of what is authorized by law. When coordinating and advancing reforms, it is necessary to pay attention to their relevance and coupling, to grasp the overall situation, and to strive to maximize overall benefits. While advancing the land administration system, we must design the systems and policies thoroughly and make good overall planning and arrangements, under the central Committee's unified arrangements and in accordance with the requirements of correct, accurate, orderly, and coordinated reform. Upholding the premise of adhering to the public ownership of land and to the use control system that centers on farmland protection, through the reform of the land management system, we shall let the market fulfill its decisive role in allocating land resources, grant farmers more property rights, improve the land expropriation system, and properly handle interest relationships relating to land, in order to modernize China's land administration system and land administration capabilities.

Second, solve the problem of convergence between reality on the one hand and policy and law on the other. For more than 30 years of reform, China has basically formed a set of effective reform strategies. That is, reforms are generally initiated from the bottom. Since existing policies had suppressed the enthusiasm of farmers, they broke through existing policy arrangements through self-initiated action, and they benefit from these policy breakthroughs. Experiments at the grass roots provide stimulus and options for reforms of current policy. Through the combination of the various forces promoting reform, consensus on policy reform is reached and the will to make a decision and act on it is formed. As these policies are promoted, the effects of reform become apparent, the policy system is perfected, and ultimately there is a push to amend the law. This is the basic experience and logic to the success of China's reforms.

The program of land system reform that took shape in the Third Plenary Session of the 18th CPC Central Committee is a positive response to the accumulated problems relating to land and to the attempts at various changes made at the grass roots. Allowing collective commercial land to enter the market with the same appertaining rights, and on an equal basis to, state-owned land is a policy recognition of the fact that a large amount of collective construction land in coastal regions and in urban-rural fringes have indeed already entered the market. The establishment of reasonable, standardized, and diversified protective mechanism is a recognition of the experiments on land expropriation that took place in a number of places. Separating farmers' contracting rights from management rights, allowing land contractual management rights to be mortgaged and guaranteed, allowing farmhouses to be mortgaged, guaranteed, and transferred, making innovations on agricultural management systems, and so on, represent outcomes of a number of local trials that rose to the level of policy. Establishing a reasonable distribution mechanism for differential income from land for the state, collectives, and individuals was the summation of local experiments and international experience. There are many more such examples.

Now that the Third Plenary Session of the 18th CPC Central Committee has designed a comprehensive plan for land policy reform, the next task—and a more difficult one—is launching the revision of the relevant laws. Otherwise, the effectiveness of these land policy reforms will diminish. The first would be the *Constitution* and how it treats the dual ownership structure wherein urban land is state-owned and rural land is collectively owned. If this ownership arrangement is maintained, then it will be difficult for collective construction land and state-owned land to enter the market on an equal basis and with the same appertaining rights. The second would be amendments to the *Property Law* and *Rural Land Contract Law*. The main provisions to be amended would: (1) modify the contract term limits and give farmers the right to use the land "over a long term without change" (*chang jiu bu bian* 长久不变); (2) separate the land contracting right from the management right and make each right fuller and more complete; (3) protect farmers' usufruct on homesteads and reform the homestead system, make the homestead and homestead ownership separable, and make homestead use rights and house rights a single, unified arrangement; (4) attach the rights of mortgage and guarantee to the land management right and to farmers' housing. The third would be to amend the *Law on Land Administration* to: (1) allow rural collective commercial land to enter the marketplace on an equal basis, and with the same appertaining rights, as state-owned land; (2) establish a reasonable, standardized, diverse protective mechanism that grants partial land development rights to land-expropriated farmers; (3) improve the secondary markets for the mortgaging, leasing, and transfer of land; (4) reduce the amount of land allocated for the public benefit and expand the scope of paid use. Of course, this list is by no means complete, but only after the amendment of these laws is completed can the plan to reform land policy be implemented and achieve salutary institutional effects.

Third, give full play to the role of experiments in formulating reform plans. One of the important items of wisdom behind China's reforms consists of running experiments to test controversial reforms in advance, and waiting to see how well they perform, and then, after the plans mature, elevating them to the policy-making level and applying them to the entire situation, so as to promote a comprehensive reform. In recent years, the Central Committee, relevant ministries and commissions, and some local governments have conducted extensive experiments in land system reform. It is recommended that the central government organize experts in related fields to carefully summarize the various reform experiments. Results from experiments on more mature reform plans would then be elevated to national policy and implemented nationwide. As for unsuccessful reform plans, plans that go against the original spirit of reform, and plans that fly the banner of reform when they are in fact merely exercises in fraud, the testing should stop. In addition, considering the controversy over land system reform and the large number of stakeholders involved, the central government may consider clarifying a package of reform trials when planning to implement the next phase of land system reform. These trials would determine such things as: how existing collective construction land can enter the market; what form the usufruct on rural households' homestead will actually have; how ownership, contracting rights, and operating rights on collective land will be separated; what the rules of the land trading market will be; what would be a reasonable distribution of land value-added revenue; how land property taxes will be levied, and at what rate, etc.

www.ingramcontent.com/pod-product-compliance
Ingram Content Group UK Ltd.
Pitfield, Milton Keynes, MK11 3LW, UK
UKHW020840290926
15470UKWH00004B/1455

9 781433 184666